Underwater Inspection

Plymouth Ocean Projects Ltd. operates at

Fort Bovisand Underwater Centre
Plymouth, Devon
PL9 0AB

Plymouth (0752) 408021

Underwater Inspection

MEL BAYLISS, DAVID SHORT
and
MARY BAX

London
E. & F. N. Spon

First published in 1988 by
E. & F. N. Spon Ltd
11 New Fetter Lane, London EC4P 4EE

©1988 Plymouth Ocean Projects Ltd

Printed in Great Britain by St Edmundsbury Press Ltd
Bury St Edmunds, Suffolk

ISBN 0 419 13540 5 (Hb)

British Library Cataloguing in Publication Data

Short, D.
    Underwater inspection
    1. Offshore structures—Inspection
    I. Title   II. Bayliss, M.   III. Bax, M.
    627'.98   TC1665

ISBN 0–419–13540–5

Contents

1 The offshore industry

1.1 PETROLEUM AND THE PETROLEUM INDUSTRY

Over the past 30 years there has been a dramatic change both in the nature of the petroleum industry and in the role of petroleum. Now almost every individual, as a consumer, is involved with oil or its derivatives. Every manufacturer in industry or in commerce, every scientist and technologist and those connected with government and politics, all have an interest in the subject.

1.1.1 Petroleum

Petroleum consists mainly of hydrocarbons; that is, a series of molecules containing atoms of hydrogen and carbon. These substances range from very light gases to solids. Petroleum is normally found at considerable depths beneath the earth's surface, where, under pressure, it is essentially a liquid. At the surface, when the pressure has dropped to atmospheric pressure, the natural gas comes out of solution according to Henry's law and the petroleum then comprises both crude oil and methane gas.

1.1.2 Natural gas

Natural gas is found in similar circumstances to crude oil and may be regarded as comprising the lightest end of the range of hydrocarbons contained in petroleum. The gas may be found in association with oil, in which case it is referred to as an 'associated gas', or it may occur on its own. When it is on its own, it may be wet gas containing certain of the lighter liquid hydrocarbons or natural gas liquids which will be separated out once the gas has reached ground level, or it may be a dry gas devoid of such liquids.

As produced it can be used directly as fuel. The heavier and more valuable gases may be separated out as liquefied petroleum gases (LPG). The great advantage of natural gas is that it is ready to burn. Its great disadvantage is that, in the absence of local markets, it is much bulkier than oil to move and to store. As a result, much natural gas in remote gas or oil fields has had to be wasted in the past – flared off when it could not be used by local consumers, as a works fuel in the fields or local refineries, or for reinjection to maintain pressures underground.

1.1.3   Crude oil

Because crude oil contains such a wide range of hydrocarbons from the lightest to the heaviest, the characteristics of individual crude oils vary greatly.

This lack of homogeneity has in fact proved one of the greatest strengths of the oil industry – in that, by choosing a suitable variety of crude oils for processing in a refinery, a close correspondence between the natural yield, distillation of the crude oil and the requirements of a market can be met.

1.1.4   Products from petroleum and their markets

The products from petroleum are as follows:

(a) Those that explode in combination with air and are used in prime movers such as the internal-combustion engine – spark or compression ignition – in road vehicles, ships or aircraft, or in static engines to drive machinery.

(b) Those that burn and provide space or process heat and light or are transformed into secondary fuels such as electricity. All liquid and gaseous products can be used for this purpose but the gases, kerosines, gas oils and fuel oils predominate.

(c) Those hydrocarbons whose combustion characteristics are poor but can be used valuably as lubricant, asphalt for roads or roofing, propellants for sprays, solvents, waxes, etc.

(d) Non-hydrocarbon or waste elements such as sulphur, vanadium, acid sludge, etc., which have to be removed as impurities from the finished products but may warrant recovery if their value in the market exceeds the cost of doing so.

(e) Feedstocks for the manufacture of further products in conversion to gas, in chemical manufacturing and the growth of protein. From petroleum chemicals are derived fertilizers, insecticides, synthetic fibres and rubbers, plastics, etc.

Although the last category of products may in the long run prove to be for petroleum the 'noblest of all', it is the first two which still predominate in volume among the wide range currently manufactured and go to meet the already large and still expanding energy needs.

1.1.5   Properties of petroleum

All fossil fuels are ultimately burned as gases. The hydrocarbons contained in petroleum are in the main highly inflammable and are contained in a compact liquid. It is crude petroleum's liquid nature which to a great extent accounts for the extreme competitiveness of petroleum in the energy field.

The fact that crude oil is a liquid permits it to migrate underground. Otherwise it would not have accumulated in commercial quantities. However, this sets explorers for petroleum a considerable problem in locating it. They can apply techniques to identify geological structures underground which may contain oil, but their chances of discovering producible oil are only about one in thirty. In other words, one 'wildcat' well in thirty drilled worldwide finds commercial oil. This means that exploration is an expensive process, as is the installation of production facilities. However, once these production facilities have been installed, the fact that oil is a liquid under considerable pressure underground makes the actual production process relatively straightforward – and much more so than for solid or near-solid fossil fuels such as coal, oil sand or shales.

1.2   CAPITAL INTENSITY AND INTEGRATION IN PETROLEUM EXPLORATION

The equipment required at each stage of the industry's operation is extremely expensive. Not only does the industry have to employ bulky hardware at each stage of its operation from oil reservoir to customer's tank, but also the oil industry is, along with nuclear energy and the chemical industry, one of the most scientifically and technologically based of any large-scale enterprise. With the delay between planning and completion of projects, this calls for the injection of large sums of money long before any income can be derived from the investment. It also means that the cost of plant (fixed costs) is high in comparison with running costs (variable costs). When these factors are linked to the high and sustained growth experienced almost without interruption until the mid-1970s – a doubling of demand and output every ten years worldwide – it is easy to see why companies involved in the business have to undertake forecasts and plan well ahead for an uncertain future.

1.2.1   Exploratory drilling

Location of the drilling site is chosen in relation to the geological picture that has been obtained. The sequence of rocks expected in the borehole is inferred from the stratigraphic and geophysical evidence, and compiled in the form of a 'prognosis'. On the basis of this prognosis, the borehole is designed. When drilling has started geological data can be compiled. Modern drilling is very destructive of the rocks penetrated, so the fragments that reach the surface in the drilling mud are incomplete and difficult to interpret. For example, clay in the rock tends to become part of the mud and be lost, while fragments of different shape, size and density travel upwards with the mud at different speeds and so arrive at different times.

The drill cuttings are sampled at regular intervals and studied in the laboratory in much the same way as the surface samples. Palaeontologists examine them for microfossils and compare the sequences and assemblage with those from surface surveys and other boreholes; sedimentologists examine the fragments, and geochemists analyse them for the organic content and any traces of petroleum. When all these data are compiled, along with depths and the rate of drilling at these depths, a fair idea of the stratigraphic sequence can be reconstructed. If the bit penetrates rocks with oil or gas, traces will also be brought up in the mud. These are detected by automatic analytical equipment installed on all modern rigs.

The logging of a discovery well provides data on the thickness and depth of the reservoirs, the thickness of oil and gas columns, porosities and estimates of the proportion of pore space occupied by water in the petroleum reservoirs. This information, with the original structural contour maps, leads to the first estimate of the total amount of oil and/or gas in place. Not all of this petroleum is recoverable with present technology, but experience will indicate the recovery factor, which may be about 20 or 25%. From all these figures, along with transport and other costs, an economic assessment is made to determine whether the accumulation could be commercially developed. But the decision to develop a petroleum field is not made on the basis of the discovery well. First comes the decision to drill appraisal wells, to determine more reliably the size of the accumulation, and to test the well contents.

1.3   THE ADVENT OF OPEC

By 1960, 45% of the world's oil outside the USA and the USSR was produced in the Middle East and over 50% of that internationally traded was exported from that area. Although the ultimate ownership of petroleum resources in all Eastern Hemisphere countries rested with the State, in one form or other, concession agreements permitted companies to establish their own development programmes, levels of production and export, and critically, their own pricing policy.

'Posted prices' had been published in the Middle East since the early 1950s and for many years these represented the arm's length prices at which crude oil was in fact available for sale on the open market. The bulk of oil produced did not, however, enter this market, but was transferred within or between the major producing companies. These posted prices were also the basis on which royalties and taxes to producing governments were calculated – the 50/50 sharing of profits was also established in the early 1950s. In many cases, their earnings from petroleum constituted the main part of the countries' national income and certainly represented the big funds on which economic expansion and social improvement could be undertaken. Such an arrangement proved feasible as long as posted prices held up or increased. With the entry of new, and re-entry of established producers after the Suez Canal closure in 1956/57, a surplus of production potential built up despite the strong growth in demand. The price at which crude oil could be sold fell. Companies' costs remained the same, realizations fell and margins were squeezed. In August 1960, posted prices were reduced in the Gulf to reflect this change in the market. Producers faced with a reduction in their revenues reacted promptly.

In September 1960, the Organization of Petroleum Exporting Countries (OPEC) was formed and the process started whereby the strength and commercial freedom of the oil companies was steadily eroded. This process reached its climax in October 1973, when effective ownership of production was transferred from operating companies to governments or their national oil companies, and with it an end to control over the development, production and price of the bulk of the world's internationally traded oil by companies operating in the OPEC area. In a matter of one decade, a major change had taken place in the relative power of producer, consumer and oil company which was fundamentally to affect the future role of petroleum, the concern of governments worldwide and the nature of oil companies.

1.4   A CURRENT VIEW OF FUTURE SUPPLY AND DEMAND

Petroleum supplies are now known to be limited and these limits have been identified. There is no longer any guarantee that internationally traded supplies will be made available merely because a market exists for them, even when the price is attractive and the cost of the basic raw material – in common with alternative sources of energy – is many times higher than the consumer and importer have been used to.

During the late 1980s or 1990s, the production of petroleum is expected to level off or even decline. In the face of these factors, and despite the still persistent tendency for energy demand to grow, the pace of that growth seems bound to slow down.

Major economies could be achieved by changes in social habits, but these, although dramatic in their effect, are unlikely to take place quickly or on a large scale. Price rises on their own may not be sufficient to induce them. Enthusiastic conservationists may walk or cycle rather than drive, but as long as they are the owners of motor vehicles they are likely to use them. Urban dwelling and improved public transport may slowly replace private commuting from dormitory towns and suburbs, but only at great cost in the quality of life and at the cost of providing both adequate transport services and the rebuilding of the centres of cities.

All these factors are likely to slow down the growth in the domestic and commercial demand for energy and oil. In the case of industry, the same elements are likely to be of even greater importance. Before the rise in the cost of fuel, the contribution of fuel costs was in most manufacturing processes relatively small in relation to total costs. Only in energy-intensive industries such as iron-smelting and steel and aluminium manufacture were they at all significant. This is no longer the case, and already there are clear indications of greater concern with the efficiency of engine and burner operation, with regular and thorough maintenance, with insulation of buildings, pipes, etc. and with the use of what was previously regarded as waste heat.

Increases in demand through to 1985 have been as low as 3% per annum – below the rate of energy as a whole and about half the rate experienced for oil in the decade from 1965 to 1975. This simple fact carries with it two basic implications. Firstly, all traditional forms of energy have to be expanded as rapidly as is feasible if energy needs are to be met. In the case of solid fuels, this not only means an acceleration of development, but means the reversal of the continuous decline in the industry over the past 25 years. Secondly, in the case of oil, because of the likely restraints upon production and exports exercised by the member countries of OPEC, all commercial discoveries outside OPEC will have to be developed as rapidly as possible. This will lead to an easing of the trading situation.

1.4.1   Consuming areas of the world

In 1976, non-communist world oil demand equalled its previous peak level of 1973, thus recovering all of the 6.5% cumulative decline of volume which occurred in the two years following the October 1973 oil crisis.

This increase in oil demand reflected the strong economic recovery many countries made during 1976, but is considered to be exceptional. In the next few years, oil demand is expected to grow on average at a considerably slower rate, probably at only about half the 7% plus rate of increase which was typical in the two decades up to 1973 – when total demand doubled every decade.

In the longer term, the continued growth in oil consumption is not expected to be sustainable much beyond 1990. Because of the finite nature of oil reserves, oil production capability is considered likely to peak out, thus imposing a limit on the oil available for consumption. Thereafter, as consumption of oil declines, other forms of energy will have to substitute increasingly for oil in competitive uses.

Most of the world's proven oil reserves are located in areas remote from the main centres of consumption. This is true even of countries such as the USSR and USA, which are the only two major consumers capable of supplying a large part of their requirements from within their own borders. Apart from the USA and USSR, no other major consuming country at present meets more than a small proportion of its needs from indigenous production. The oil industry has to equalize this imbalance between demand and supply by moving enormous quantities of oil, along routes up to 12 000 miles long, from surplus to deficit areas. This is a large and complex task.

The biggest single oil-importing region is Western Europe. Over half of total European energy requirements are met by oil and until the arrival of North Sea production, all but about 5% of this oil was imported.

Outside Europe, Japan produces virtually no oil of her own and depends on imported oil for about three quarters of her total energy supplies. The high rate of growth of European oil imports between 1963 and 1973 was dwarfed by the surge of imports into Japan, which, over the same period, increased by an annual average of nearly 17%. In tonnage terms, Japan's imports increased fivefold from 62 million tonnes a year in 1963 to 284 million tonnes a year in 1973, and Japan has continued to have a very heavy dependence on imported oil, especially fuel oil.

A recent development over the past 15 years has been the strong growth of oil imports by the USA − essentially a reaction to the peaking out of domestic oil production, and delays in the development of new oil and other indigenous energy resources. The USA has long been a substantial importer of oil from Western Hemisphere sources in the Caribbean. What is new is the massive increase in imports of oil from the Eastern Hemisphere sources of Africa and the Middle East, in some cases after processing at offshore refineries in the Caribbean. Concern in the USA about this increasing dependence on Eastern Hemisphere imports has been an important element in the political and economic decisions which have dramatically changed the industry in the years since then, and especially since the American hostage incident in Iran between 1979 and 1981, when patriotic factors as well as economic ones were major items of consideration.

However, the oil industry has to do more than just match the volume of oil required in a particular country with an equivalent supply. It has to reconcile, at lowest cost, both the product pattern and quality requirements of the consuming areas with the demands in terms of price and political conditions in the countries which supply the oil.

The pattern of oil consumption differs widely from area to area and often from country to country within an area. In the USA, for example, the dominance of the automobile, the high production cost of domestic crude oil, the availability of indigenous natural gas and cheap coal, have combined to require the oil industry to convert into motor spirit nearly one half of every barrel of crude oil it refines. At the other extreme, in Japan the lack of alternative energy sources and Japan's quite understandable aversion to the principle of nuclear power, have resulted in a market demand heavily weighted towards fuel oil.

Seasonal variations are also important with regard to the total world oil supply pattern. In most developed countries, the peak demand for oil is during the winter months. Because of world geography, the peak world consumption is during the Northern Hemisphere winter.

Future worldwide oil consumption levels will depend on the interplay of several factors including general economic growth, and on how quickly other energy sources are developed. There is likely to be some substitution for oil in certain markets by natural gas and also by coal and nuclear power. Linked with this substitution by other fuels, we are likely to see a gradual lightening of the barrel and the eventual limiting of the use of valuable petroleum fuels to mobile transport, where no easy substitute is in prospect, and as feedstock for chemicals, plastics and proteins.

1.4.2   Changing patterns of supply

We have also begun to see some change in the pattern of oil supply as well as demand, as new areas of production, notably the North Sea and Alaska, have made a contribution to oil supplies. Oil from the North Sea is now flowing in quantity and production has built up so that the United Kingdom has achieved net self-sufficiency. North Sea oil has also had an important effect on the total European oil supply pattern, although there is still considerable dependence on the Middle East, which in world terms remains pre-eminent as a production area. It is possible that Japan could take increasing supplies of oil from China and/or Russia, while the USA and Canada, which have been increasing their imports of oil especially from the Eastern Hemisphere over recent years, may at least stabilize their levels of imports after the starting up of oil production from the North Slope of Alaska in 1977, and exploration and development off the coasts of Labrador and Newfoundland in the early 1980s.

2 Offshore installations

For well over a century, oilmen have been trying to locate the places where oil or gas lies under the ground. Latterly, the search has been extended to include the rock formations under the sea.

The first activities to precede any permanent installation are those of the geologist and the geophysicist. Their work is to survey rock formations beneath the sea by any means possible, but nearly always by using ships. Geophysicists set off explosions from their ships and measure the reflected waves by hydrophone sensors. Geologists use their knowledge of the surrounding land, obtained by drilling core holes, and fill in the gaps in knowledge with information from magnetometers and gravimeters.

Once the location of an oil or gas deposit is established and a detailed survey has located the best possible place for a well, then drilling can begin. The cost of such an operation on shore is high, but offshore it is much more expensive, a deep well taking several months to drill and costing millions of pounds.

2.1 EXPLORATION WELL DRILLING

Drilling rigs for offshore exploration take several forms. Three of the most common types are:

(a) The jack-up platform.
(b) The semi-submersible platform.
(c) The drillship.

2.1.1 Jack-up platforms

The jack-up platform (Fig. 2.1) consists of a main platform body or hull attached to which are legs which can be lowered to the seabed so that the drilling platform can be raised above the water for drilling, or raised up so that the platform can be towed away to a new site.

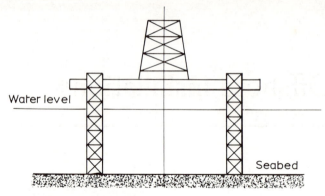

Fig. 2.1   The jack-up platform.

2.1.2   Semi-submersible platforms

In deeper water, the legs of a jack-up platform would be so long as to give concern about the stability of the legs unless they were much larger in section. Therefore, instead of jack-up legs, the semi-submersible platform (Fig. 2.2) has large hollow floats, and this type of platform floats with a large part of itself submerged, rather than standing on the seabed. This flotation is achieved by flooding the floats. When moving the platform from one site to another, most of the water is pumped out of the floats and the platform is then towed away. When drilling, the semi-submersible platform is held in position by anchors, and where the water is too deep for anchors, propellers keep the platform stationary. These propellers are controlled by computer, which compensates for the effects of wind and current.

2.1.3   Drillships

An alternative to deep water drilling is to use a specially designed drillship. It has the capabilities of a ship in that it can move around from site to site with the

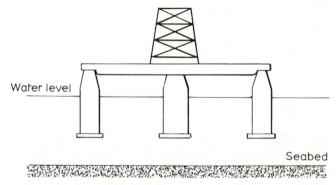

Fig. 2.2   The semi-submersible platform.

ease and speed of a ship. To hold the ship stationary while drilling on site, either anchors or dynamic positioning by propellers are used.

2.2   THE EXPLORATION WELL

The well itself is a hole drilled through the rocks. To do this, the main feature of a drilling rig is the derrick, a tower in which lengths of pipes, which form the drilling string, can be raised or lowered into the bore of the well by a winch. Generally, large diesel engines or gas turbines provide the power for the winch and the motors that drive the rotary table, a round section of the drill floor at the base of the derrick that rotates the drill string.

Once the well has been drilled some distance into the seabed, the surface section of the well has a large diameter steel pipe cemented into it. This pipe is known as the casing. The blow-out preventer is fixed to the top of this casing at seabed level. As its name suggests, the function of this piece of equipment is to stop the oil or gas blowing out of the well in an uncontrolled manner in the event of an emergency. It is powered by very powerful hydraulic actuators that can seal off the well if the pressure rises to a dangerous level.

The drillpipe is lowered into the hole under its own weight and is turned by the rotary table. The weight and rotation of the drillpipe cut their way through the rock. The top section of the drillpipe is square rather than round in section, and this is used to rotate the drillpipe. This part of the drillpipe is called a kelly. When a new length of drillpipe has to be added, the kelly is unscrewed, the new length of round pipe is attached to the drillpipe, and the kelly is screwed on again. Drilling is then restarted.

The speed of drilling depends on the hardness of the rock. There are different designs of drill bit. Some of these work on the actions of small tools rolling over the surface, while for hard rock, diamonds are embedded into the surface of the bit.

In order to remove the drilling debris and so keep the tool from clogging up, a fluid slurry is used, which is referred to as drilling mud. The drilling mud also keeps the bit cool and lubricates it, as well as providing a counterbalancing pressure to the pressure of water, oil or gas found in the various rock strata. The mud is pumped down through the hollow drillpipe and returns through the space between the drillpipe and the well casing. See Fig. 2.3.

2.3   PRODUCING OIL AND GAS

Of the thousands of exploration wells drilled, only about one in thirty finds oil or gas. If nothing is found, the hole is sealed with cement. When oil or gas is found and tests show that the well is likely to produce a steady flow, the next

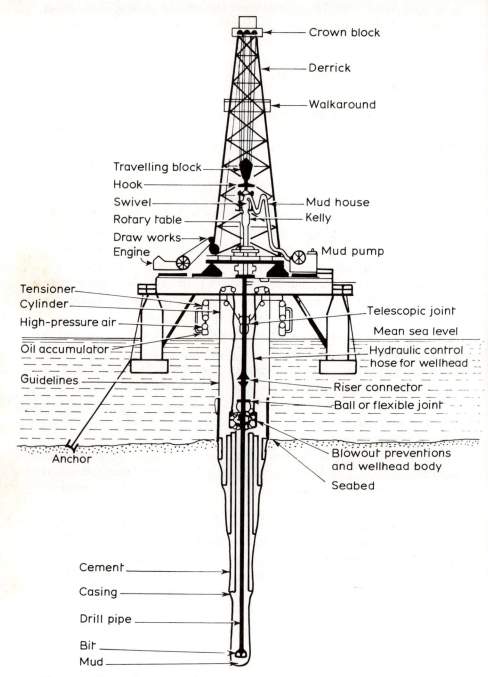

Fig. 2.3 The exploration well.

task is to line the hole with a steel tube, the case, down to the oil- or gas-bearing rock, which is often referred to as the pay zone. Tubing is now lowered into the well to a point opposite the pay zone. This production string is held in place by a tubing hanger at the top of the hole. At the bottom of this tubing, safety valves are mounted so that they can be shut off in an emergency.

At the top of the well, an array of valves is mounted. This is known as the Christmas tree. These valves allow control of the well flow from its various zones (if it has more than one pay zone) and allow access into the well, as well as directing the flow once it comes out of the well.

Pipelines are run from all the wells on the seabed to the surface of the sea. These are collected together on installations that house all the equipment for handling the oil as well as living accommodation. These offshore platforms are artificial islands from which the oil production from the seabed can be managed. The world's biggest platform measures just over 304.8 m (1000 ft) from the seabed to the surface. These platforms may be made of either steel or concrete.

2.3.1   Steel platforms

If the platform is made of steel, the structure on which the installation is supported is known as a jacket, because it acts as a jacket around the wells. The jacket is built in a fabrication yard and is either floated out to the site or carried out on a barge. During the fabrication of the jacket, thorough inspection is required of all the welded joints, especially the nodes.

When the ocean site of the platform is reached, the structure is lowered to stand on the seabed, and anchored. The piles that anchor the structure to the seabed are sometimes driven several hundred feet into the seabed. The deck is then lifted onto the jacket, which forms the platform on which are mounted all the equipment, accommodation cabins, helicopter platform and other facilities required to produce oil or gas from the field. The design of each of these items is arranged to fit in rectangular boxes called modules, each one being as completely fitted out as possible on land so that once installed on the structure all that is required is the connection of services to and from each module. The biggest module in volume is usually the living accommodation, and the heaviest (for example, a couple of thousand tons) a complete diesel-driven power station.

2.3.2   Concrete platforms

Concrete platforms are also built ashore. They are invariably free-standing. Their enormous weight holds them stationary on the seabed, without the use of piles. The large honeycomb-like structure of the base means that they can be floated out to the production site and then sunk by flooding the honeycomb structure at the base. The deck and modules are placed on the concrete structure inshore, usually in the partially sunk position, the honeycomb base structure being pumped out before being towed to the oil field.

2.3.3   Installation systems

On land, it is possible to drill several different wells at great distances apart and connect them to a central collection point. At sea, however, all the wells are drilled from the platform. The technique of angle drilling allows each well to be angled below the seabed so that it reaches a different part of the oil field. A platform may act as a collection centre for as many as sixty wells.

As the production from larger gas and oil fields becomes necessary, a change in the system of installation is noticed, namely, the individual well with a seabed wellhead assembly. Pipelines then carry the oil or gas to a nearby platform. In this way, it is possible for one platform to produce from a larger area than can be achieved with angled wells drilled from the platform.

Development is currently going on to produce a system whereby the wellhead and processing equipment can be installed on the seabed. However, it will be some time before the production platform disappears in favour of a complete subsea installation.

Once produced, the oil has to be distributed to the refineries or petrochemical plants as soon as possible for processing. This is done in one of two ways, either using pipelines which bring the crude oil straight to the land, or using tankers that load oil at the oil field and discharge it at terminals near or at the processing plant.

2.4   PIPELINES

In the very early days, oil was carried in barrels. Hence, the traditional measurement of 'barrels of oil'. Wagon owners thought they could charge high transport rates forever, so the oilmen's thoughts turned to pipelines as an alternative. At sea, specialized ships for carrying oil, tankers, started to appear at the end of the last century.

Oil and gas pipelines are almost always made of steel capable of taking the high pressures, but plastic pipes are now being considered seriously for their light weight, ease of construction and maintenance.

Gathering pipelines in an oil field may be only metres long and millimetres in diameter, whereas the major supply lines run across many countries and for thousands of kilometres under the sea, and are large enough for a man to walk upright in them.

The advantage of a pipeline is that it is generally cheaper to transport oil by this means than by road or rail. Large tankers are marginally cheaper, but the pipeline is more convenient in that it is not subject to weather or industrial uncertainties. Shore capacity and the provision for extra branching can be designed into a pipeline system, which then gives flexibility in future distribution.

Pipeline construction for oil and gas is generally carried out by specialist companies. The pipes are often laid in trenches on the seabed and buried in order to protect them from damage such as an anchor being dragged across them.

Pumping stations are sometimes required to keep the oil flowing, although for short pipe runs there is sufficient pressure in the oil itself to move it along. Now, except for short pipelines, pipelines are constructed and laid from special barges. The pipe is welded, coated with bitumen, and sprayed with concrete to give it weight as well as to protect the steel from damage. The extra weight is to prevent the pipe from floating off the seabed. The completed pipe is lowered into the water as the barge moves along the proposed route of the pipeline. Pipelines are protected from corrosion by coating and one of the forms of cathodic protection.

The pipeline is inspected and cleaned by devices known as pigs or go-devils. The device is pushed through by the oil and scrapes away the wax and dirt from the walls of the pipe. Pigs are used to separate different products being pushed along a pipe one after the other. Sometimes they are used with cleaning fluid, but at other times they are used to separate different types of crude oil. Often, the pig will carry instruments for internal inspection of the pipe.

3 Regulations relating to inspection

3.1 REASONS FOR GOVERNMENT LEGISLATION

Legislation concerning inspection requirements has primarily dealt with fixed installations and in particular the 'fitness' of the structure itself. Government authorities have had less cause to set out regulations for the inspection and maintenance of production equipment *per se* since there are considerable commercial incentives for the operator to ensure optimal continuous operation of production items despite the potential conflict between short and long-term objectives. Furthermore, a considerable proportion of production equipment is located on the platform and is therefore proportionately cheaper to inspect and repair than underwater structures and pipelines.

An additional reason for the preoccupation of state agencies with structural and hydrocarbon transfer/transport aspects lies in the motivation and expressed reasons for governmental regulations. National authorities are primarily concerned with maintaining maximum safety levels in respect of personnel, the environment, and marine life. However, it would be naïve to ignore secondary objectives which may be realized by national rules and standards governing a wide range of offshore exploratory, developmental and production activities, such as increased demand for local labour and resources and increased state participation and control.

3.2 INCREASE IN INSPECTION REGULATIONS

Legislation governing inspection and maintenance requirements for offshore installations is reaching its most advanced stages at present in the north European continental shelf.

3.3   REGULATIONS IN NORTHERN EUROPE

The north European continental shelf is divided into separate national sectors for the purposes of mineral and hydrocarbon exploitation. In most of the sectors there is a government requirement that fixed installations have a certificate of fitness, the primary purpose of issuing the certificate being to ensure the safety of personnel on the structure. Retention of the certificate requires extensive work to ensure the continuing safety of the installation.

In addition to the requirement for safety of personnel, the operator requires the installation to remain efficient and capable of continuous production. This involves maintaining the production equipment in good operational condition, and while most of it is situated on the topside of an installation, there are some parts under water which need to be monitored.

3.3.1   British regulations

In late 1965, the jack-up barge Sea Gem collapsed with the loss of 13 lives. A subsequent enquiry led to the *Mineral Workings (Offshore Installations) Act 1971*, which authorized the Secretary of State to implement regulations for the safety of offshore installations. In 1974, the Offshore Installations (Construction and Survey) Regulations were issued under the Act, laying down regulations on the standard of design and construction of a platform. The regulations also outlined survey requirements necessary to maintain certification of an installation through its offshore life.

The Department of Energy (DEn) then issued a guide to the technical standards required to satisfy the regulations. This guide gives an indication of the minimum standards to be accepted when an application for a certificate of fitness is considered. It is not a legal document. A second edition of the guidance notes was published in December 1977.

The Department of Energy has named six organizations which can carry out certification on its behalf. They are: the American Bureau of Shipping, Bureau Veritas, Det norske Veritas, Germanischer Lloyd, Halcrow Ewbank Offshore Certification Bureau, and Lloyds Register of Shipping.

3.3.2   Norwegian regulations

The Norwegian requirements are similar to those of the British, but there are differences in emphasis. The Norwegian government, through its Petroleum Directorate, employs an independent agency to carry out surveys and other recertification work on its behalf, and this has usually been Det norske Veritas (DnV). The legislation giving the requirements for underwater inspection is given in the Royal Decree of 9 July 1976. The first draft of guidelines for inspection was circulated for comment in April 1977.

3.3.3   Dutch regulations

The Netherlands issued regulations in 1967 based on their *Continental Shelf Mining Act 1965.* These are the Mining Regulations (Continental Shelf) and are regularly updated.

The regulations require a certificate of construction which includes a check of the design, approval of construction to the required standards and successful installation. The final approval is at the discretion of the Inspector General of Mines (IGM). Regulations covering regular inspection and renewal of certificates at specified intervals are being prepared. For the certification of construction the IGM approves five of the certification societies approved in the UK (i.e. all except Halcrow Ewbank Offshore Certification Bureau).

3.3.4   West German and Danish regulations

Both West Germany and Denmark have produced legislation concerning certification of offshore installations but few details are available in English. It is understood that the Danish ones closely follow the requirements of the certifying authority Det norske Veritas.

Statutory regulations on the inspection of offshore installations do not lay down definitive inspection requirements for any platform. Each installation is designed for a specific location and use, and it follows that the inspection requirements for each platform vary accordingly. Regulations therefore generally lay down broad survey requirements for maintenance of certification and require that, for each installation, the inspection schedule should be agreed between the owner and appointed certifying authority.

The following statement appears in the guidance notes in the section on Major Surveys: Recertification:

'In a jacket-type structure, the outer more accessible, members may be taken as representative of the internal members unless there is evidence to suggest otherwise.'

It is for the owner and certifying authority to agree which areas on a particular structure can be taken as representative.

A part of the section on objectives states that:

'. . . an agreed schedule of inspections and tests should be drawn up . . . [which] should be arranged by owners and Certifying Authority and, where appropriate, in consultation with the organisation responsible for the underwater inspection and testing and any other specialist body whose services will be required.'

In this context, the design organization is a 'specialist body' and must play an important role in the planning of inspection details.

3.4 CERTIFYING AUTHORITIES

Six certifying authorities at present operate on the north European continental shelf; i.e. all those approved by the British Government. Five of them are classification societies approved by the various governments to act as certifying authorities on their behalf. The sixth, Halcrow Ewbank Offshore Certification Bureau, operates as a certifying authority only in the UK sector. The Netherlands approve five classification societies for the certificate of construction and it is understood that they will do the same for recertification work. In the Norwegian sector, DnV and Lloyds Register of Shipping do consultancy work for the Norwegian Petroleum Directorate, but the Norwegian Government remains the sole certifying authority. The governments of other sectors have not yet made firm policies on recertification work and have not appointed any certifying authorities.

3.4.1 American Bureau of Shipping

The American Bureau of Shipping is currently producing regulations which will closely follow the DEn guidance notes.

3.4.2 Bureau Veritas

The French classification society, Bureau Veritas, has published a set of regulations for classification of offshore installations. These are the Rules and Regulations for the Construction and Classification of Offshore Platforms. Although they are for classification purposes, they can also be applied to the granting of a certificate of fitness. The underwater inspection requirements are as follows:

(a) The owner must apply for surveys as necessary, and provide full facility for the survey.
(b) Special surveys must be undertaken every four years, but an extra year may be allowed for performing the survey. There must be a detailed examination of essential parts of the structure, especially columns, supports, bracings and articulations. Non-destructive testing may be required on parts subject to corrosion or subjected to alternating loads. All systems and devices used in the survey are to be defined in agreement with the surveyor.
(c) Annual surveys include visual inspection of the outer parts of the structure, particularly those parts near the surface, to check for corrosion. Accessible welds which are subject to variable stress must also be inspected. The annual survey may be extended to sub-surface areas.
(d) Occasional surveys are to be carried out after modifications.

These rules are left open to interpretation in several areas, and so particular care should be taken in agreeing the extent and detail of inspection with Bureau Veritas before detailed design of a structure is undertaken.

3.4.3   Det norske Veritas (DnV)

This is a Norwegian classification society. The main points concerning underwater inspection from the 1977 rules are as follows:

(a) The maintenance of the certificate of approval requires that the structure is subjected to periodic surveys, that it is operated in accordance with the operations manual approved by DnV, and that the owner expediently notifies DnV of conditions, events or planned actions that may make it necessary to perform a special survey. It is assumed that the owner will carry out running inspection as required to maintain the structure in a safe condition.

(b) In principle, the periodic survey is to comprise inspection of selected elements of the structures. . . . The rotational basis except for elements that are monitored regularly for trend analysis or other reasons. The extent of each periodic survey is to be based on accumulated evidence . . . [from] earlier surveys. Normally each survey is to include:

  (i) General visual inspection of selected parts of the structure to determine the general condition of the structure and to locate areas that should be subjected to close inspection and testing.

 (ii) Close visual inspection and non-destructive testing of selected local areas of the structure to detect possible material deterioration or incipient cracking.

(iii) Visual inspection and testing as needed to check the condition and function of corrosion protection systems.

(iv) Inspection as needed to check the condition of the foundation and of scour protection systems where installed.

 (v) Inspection as needed to determine the amount of marine growth on the structure and the presence of debris in contact with the structure. In conjunction with surveys, cleaning of structures to be inspected is to be carried out as necessary. The survey programme is to be scheduled so that the whole structure is covered in a period of five years.

(c) In the event of accident, discovery of damage, or deterioration that may affect the short-term safety of the structure, a special survey may be required.

The survey requirements of DnV clearly can be very extensive.

3.4.4   Germanischer Lloyd

Germanischer Lloyd are preparing regulations for the certification of offshore installations. It is understood that they will follow the guidelines set out in the DEn guidance notes.

3.4.5   Halcrow Ewbank Offshore Certification Bureau

This certifying authority, comprising Sir William Halcrow and Partners, Ewbank and Partners Ltd, Harris and Sutherland, Berness Corlett Marine Ltd, and Messrs Sandberg, was appointed by the British Government in late 1976. By late 1977, the organization had not applied for permission to act in areas outside the UK sector. The group has not issued any regulations covering underwater inspection.

3.4.6   Lloyds Register of Shipping

Lloyds Register of Shipping currently base their survey requirements on the DEn guidance notes. Rules are being prepared.

3.5   OPERATIONAL REQUIREMENTS

During the course of the development of the North Sea gas and oil fields, there have been several accidents, and partly as a result, several items of legislation have been brought about (see Fig. 3.1). Both the accidents and the legislation have influenced the development of North Sea inspection techniques.

Statutory requirements for inspection of installations are the minimum requirements to ensure the safety, health and welfare of persons working on them. In that context, they cover everything on a structure with the exception of risers, conductors, intakes, drains, etc. (i.e. the statutory requirements are not concerned with operational functions). However, the inspection requirements of certain parts of a structure may be governed more by the operation of the platform than by the safety of its personnel. This section looks first at those areas requiring inspection to ensure the continuing function of the installation and then gives some examples of failures of components affecting operations.

Inspection of risers and conductors must be considered essential. They carry oil and gas, sometimes at elevated temperatures, and may be highly susceptible to deterioration. At present, there are no British regulations covering the inspection of risers and conductors, but statutory requirements are being formulated. The Norwegians, however, do have regulations for the inspection of risers.

Risers and conductors are perhaps the most important and most publicized items requiring regular inspection. However, there are a number of other components, the performance of which can affect the operation of a platform.

Date	Event
1950	First gas exploratory wells drilled in the southern North Sea
July 1960	SI 688 1960 The Diving Operations Special Regulations
December 1965	Sea Gem collapsed
June 1972	S 1702 1972 The Offshore Installation Registration Regulations 1972
March 1974	Department of Energy guidance notes for fixed platforms
May 1974	*The Mineral Workings (Offshore Installations) Act 1971*
January 1975	SI 229 1974 The Offshore Installation (Diving Operations) Regulations 1975
March 1975	DI 116 1975 Merchant Shipping (Diving Operations) Regulations 1975
August 1975	Certificate of Fitness Offshore Structures required
July 1976	SI 923 1976 Submersible Pipelines (Diving Operations) Regulations 1976
July 1976	SI 940 1976 Merchant Shipping (Registration of Submersible Craft) Regulations 1976
1977	Major oil leak in Norwegian sector caused by faulty BOP (blow-out preventer)
1980	Alexander Kielland disaster
July 1981	Diving Operations at Work Regulations 1981

Fig. 3.1   Major events and UK North Sea legislation.

Three examples of such components are:

(a) Seawater lift intakes and caissons (excluding firewater lifts).
(b) Oil flow lines, other than risers and conductors.
(c) (Some) supply boat mooring systems.

Another factor, which may affect pipework and some topside equipment, is change in platform attitude (i.e. tilt, settlement, lateral movement).

Operational failures have occurred on platforms in the North Sea only months after their installation. The following examples illustrate the types of problem that have arisen.

In November 1976, an oil field in the northern North Sea, with three production platforms in operation, had to shut down production on two of them. Although production was resumed on one of them a few hours later, the total loss in output was substantial. Shutdown was made necessary by an oil leak on one of the production platforms due to failure of a single pipe.

On the same field, a production riser was due for replacement at the same time, because of rapid corrosion. Both these occurrences were within one year of the start of production.

A few weeks earlier, in another field, a concrete platform almost had to shut down production. A seawater intake, at a depth of 65 m (213 ft), which provided cooling water for production equipment, was blocked with marine growth. Two points arise from this occurrence. Firstly, design information had stated that deeper than 46 m (151 ft) equipment should not be in danger of marine fouling, whereas in practice a grid fouled at 20 m (66 ft) below that depth. Secondly, based on the design information, the grid could have been placed at a depth between 46 m (151 ft) and 50 m (164 ft), which would have facilitated maintenance by air-range divers. At 65 m (213 ft), any manual work would have to be carried out by mixed-gas divers.

In autumn 1975, a corroded riser on a central North Sea platform almost caused a complete shutdown of production. The riser ruptured in the splash zone while under test and was found to have corroded under its protective coating. The platform and one other in the field were shut down, cutting the total production by over 80%. This situation lasted two days, after which flow was restored to two thirds of the original. The loss of revenue to the operators is said to have reached about $2.5 m per day.

In 1980, the failure of a component on the accommodation platform Alexander Kielland led to the capsize of the platform, resulting in the death of over 150 persons.

Exploration in the North Sea is at present levelling off, and the production phase is in the middle stages of a 20 or 30 year life. Yet many of the above problems occurred during the first year of production in the fields concerned. It is clear from these examples that careful design of production systems and ease of access to critical areas are of major importance. The design must facilitate easy inspection and maintenance so that a regular check can be made on the components.

National sectors in the North Sea are given in Fig. 3.2.

3.6   THE NEED FOR INSPECTION

Underwater inspection is a continuation of the onshore fabrication of a facility through shipment, installation, to operation. It is first required during installation and then periodically when a facility is operational.

During installation, an almost continuous surveillance is carried out to ensure that the integrity of the structure is maintained, although underwater inspection during this phase is a small part of the total demand for underwater inspection.

Inspection during the operational life of the installation can be for any one of several reasons, including:

(a) Certification, or the maintenance of a certificate of fitness.
(b) Operators' assurance of reliability and safety.
(c) Work associated with accidents, repairs after accidents and other modifications.

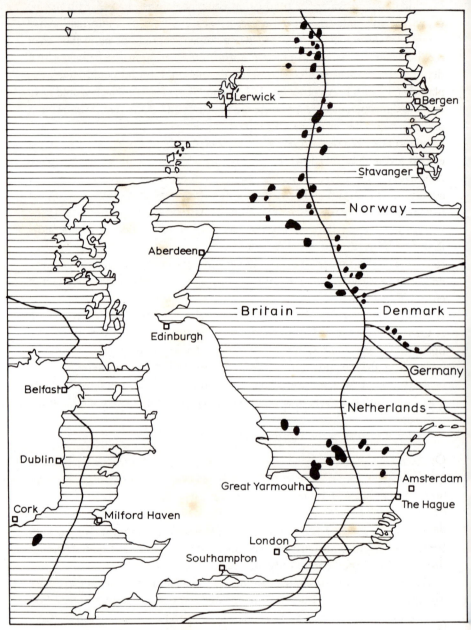

Fig. 3.2 National sectors in the North Sea.

Poor quality work at the fabrication and installation stages can lead to very expensive problems later in the life of a facility.

From discussions with the Department of Energy, it is clear that some accidents have occurred which have required underwater repair. The majority of these accidents occurred during the installation of the facilities and during the placing of modules and topside equipment. The principal causes of damage were ramming by construction supply boats, falling equipment and storm damage. Accidents not related to installation include trawls fouling seabed equipment and damaging exposed lengths of pipe. There is little quantitative evidence of their extent.

However, from the survey of operating companies, it appears that inspection associated with the repairs arising from these accidents is small compared with the total inspection effort required for certification and operational assurance.

The introduction of the UK Offshore Installations (Construction and Survey) Regulations 1974 created a backlog of inspection necessary for the issue of certificates of fitness for all platforms installed prior to 1974. The backlog has now been cleared and unless the certification regulations are changed and enacted retrospectively, no further peaks in demand for inspection work are envisaged.

Currently, the majority of inspection is related to certification and the operators' assurance of reliability and safety.

3.7   REQUIREMENTS FOR INSPECTION OF STRUCTURES

Operators' assurance of reliability and the legal requirements for certification (particularly in the British and Norwegian sectors) have created the need for underwater inspection of the structure of the installations during their operational life. Practically all operators have devised, or are in the process of devising, annual inspection programmes which in aggregate meet the five-yearly recertification requirements. This eases the inspection requirements of the individual operators.

The details of operators' programmes vary, but they are all basically similar. The general features of the annual programme include:

(a) A complete general visual survey of structure and risers.
(b) A pattern of readings on the cathodic protection system covering 10 to 20% of sacrificial anodes where this system is installed.
(c) A close visual survey of about 10% of nodes including critical nodes every year; depending on the structure, this may lead to non-destructive testing (NDT) (usually using magnetic particle inspection techniques) of certain critical nodes as dictated by certification requirements.

(d) Preparation of a scour map.
(e) A visual survey of pipelines and riser connections within a short distance
   (e.g. 20 m (66 ft)) of the platform perimeter.

On all the older platforms, non-destructive testing (NDT) inspection of a selection of critical node welds is specified. However, because of design confidence, on many of the newer platforms no NDT inspection of nodes is carried out. Current inspection effort is increased by many operators conducting information-collection exercises on their structures.

From a survey conducted amongst operating companies, it is clear that routine and certification inspection work currently accounts for about 75% of all underwater inspection work on structures. The remainder is associated with construction or repairs.

3.8   NUMBER AND TYPES OF STRUCTURE

On the north-west European continental shelf, there are currently 112 platforms installed, under construction, or ordered. Of these platforms 110 are in the North Sea, two being in the Irish Sea. In the North Sea, latitude 56° N is the approximate northern limit of sea which is 50 m (164 ft) or less in depth and therefore the limit of air diving under British regulations. Of the platforms, 62 are in these southern waters. All the remainder, of which 13 are concrete gravity structures, are in water deeper than 50 m.

In addition to the platforms, there are about 40 other structures including flare structures, buoyant towers, loading buoys and subsea completions. In the UK sector, there are 65 platforms or other installations, of which 11 are concrete. The Frigg Field is assumed to be in the UK sector.

The first installations were placed in the North Sea in about 1966, in water less than 50 m (164 ft) deep, south of 56° N. There has been steady growth in the number of platforms south of 56° N since that time, but it is now slowing as exploitation limits are reached.

The first installations in the deep water north of 56° N were placed in about 1972 and there has been rapid growth in the number of installations. Again, the rate has slowed down, in this instance because of the pattern of discoveries and resource-depletion 'policies' taking effect. Currently, the deepest platform stands in the Magnus Field. It is forecast that there will be about 140 platforms installed by 1987, with most of the additional platforms in deep water.

In such a speculative area, it is preferable that private sector industries should plan on the basis of cautious estimates of growth. A reasonably optimistic view would assume about 15 more platforms by 1987, giving a total of 155 installations.

REFERENCES

1. UEG (1978) *Underwater Inspection of Offshore Installations: Guidance for Designers* Report 10, February, Underwater Engineering Group, London.
2. UEG (1979) *The Market for Underwater Inspection of Offshore Installations in the Next Ten Years*, Report 13, February, Underwater Engineering Group, London.

4 Engineering of offshore installations

4.1 THE NEED FOR OFFSHORE STRUCTURES

In the search for oil, it is usual for the drilling to follow a set procedure. After initial exploration and reconnaissance using seismic survey techniques, specialist geologists predict a likely position for oil deposits to be found.

An exploration drilling rig is despatched to this position and it drills down into the substrata in the hope of discovering recoverable deposits of crude oil. If this rig is successful, the oil deposits are investigated by further drilling to ascertain the amount of recoverable oil in the field and thus assess the feasibility of developing these deposits commercially.

Once it has been decided to go into production on a particular field, then it becomes necessary to place a production drilling facility over the site. This structure then remains in place for the life of the field and during that time many boreholes may be drilled into the oil deposits so that they can be recovered on the surface and then shipped off to refineries on a worldwide basis. In the North Sea, these production facilities are referred to as 'platforms' and the majority of them are fabricated from steel.

4.2 STEEL STRUCTURES

The technology associated with the fabrication of steel structures has improved and enabled very large structures to be designed, fabricated and installed. It must be borne in mind that all offshore work is expensive, whether it be installation, hook-up or maintenance – using crane barges, tugs, supply boats, helicopters, diving ships, submersibles, remote controlled vehicles, etc. Thus, offshore work should be minimized by creating structures which can be easily installed and which require minimum maintenance and repair.

4.3   PILED STRUCTURES

Most offshore production platforms are of the piled steel structural type. This type of configuration is made of tubular steelwork welded into a framework with vertical or near-vertical legs. The piles are tubular steel and are driven into the seabed either through or around the main legs by pile hammers (see Fig. 4.1). Pile diameter varies between 600 mm (24 in) and 1800 mm (71 in), depending on the size of the structure and the number of legs. In general, structures have between three and eight legs, depending on the structure's height, on the seabed's load-bearing capacity, and on the mode of transportation to the structure's site — i.e. self-floating with inherent buoyancy, or barge-mounted.

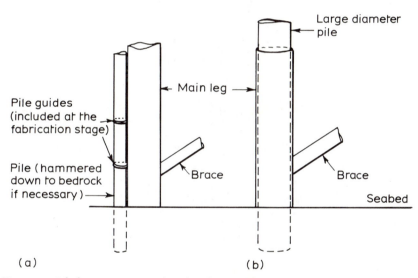

Fig. 4.1   Piled structures: (a) piles placed around main leg and hammered into the seabed; there are several piles per leg; (b) large diameter pile driven into seabed through main leg.

4.4   SELF-FLOATING STRUCTURES

These structures have flotation units to enable the structure to be towed to site. On site, the units are flooded in a pre-set sequence to right the platform and lower it to the seabed. The flotation may be provided by large diameter legs (up to 9 m (30 ft) in diameter) split into compartments which can be used for oil storage when the structure is producing (for example, the Thistle platform). The other method is for flotation tanks to be attached to the structure and removed after the installation is complete (for example, Forties Field jackets). See Fig. 4.2.

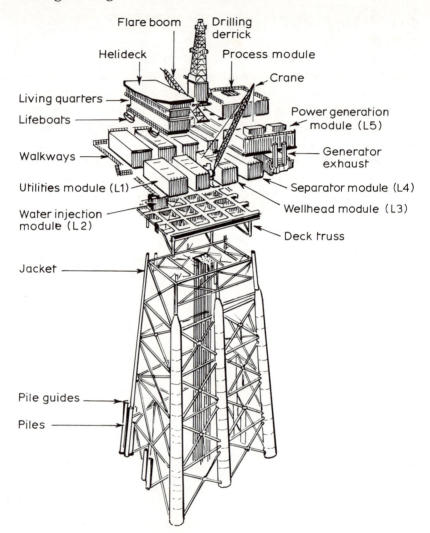

Fig. 4.2 Brent 'A' production platform.

4.5 BARGE-MOUNTED STRUCTURES

Many structures are transported on barges and do not have extra flotation incorporated into the fabrication. This reduces the leg diameters of these structures. Extra bracing may be incorporated to enable the structures to withstand the stresses of the barge launch. After the launch, the jacket is either towed into place if the structure is buoyant, or lifted into place by crane ship (for example, southern North Sea gas platforms).

4.6 TECOMARE/CHICAGO BRIDGE (STEEL) DESIGN STRUCTURES

These structures are a relatively recent development for steel fabrication, consisting of a tubular structure with a number of ballast/storage tanks. The structure is usually floated to its site vertically, the tanks being used initially to provide buoyancy and then ballasted with water. After installation, these can be used for storage of the petroleum product.

The depth of water may limit the use of this type of structure for the northern North Sea, although concrete structures of this type have already been installed here.

4.7 GRAVITY CONCRETE STRUCTURES

Before the development of large derrick barges, the installation costs of steel platforms were immense. This led to the search for a material to replace steel. The material chosen was concrete, as it was relatively cheap, easily moulded and its behaviour was well understood. Concrete structures were designed and built in direct competition to steel and until very recently had some advantages over steel, such as:

(a) Cheaper than steel for fabrication and installation, although there is a similar weight of low-grade steel in the form of reinforcing bars as there is high grade steel in a steel structure.
(b) Faster than a steel fabrication, using easily available materials.
(c) Incorporation of a storage facility.
(d) Few corrosion problems.

All the concrete structures currently in use and being built are of the gravity type. The major design differences which various companies have involve the support of the deck and the method of overcoming the environmental forces.
The major designs are:

(a) Condeep (for example, Mobil Beryl, Shell Brent 'B' and 'D') (see Fig. 4.3).
(b) Howard Doris (for example, Total MCP 01).
(c) McAlpine Sea Tank (Cormorant 'A') (see Fig. 4.4).
(d) Andoc (Dunlin A7).

4.7.1 Condeep design

This is a three-legged structure, supported by a series of storage tanks, with a steel deck. Because of the large base area of concrete structures, scouring is a problem. This is overcome by the provision of a skirt which penetrates the

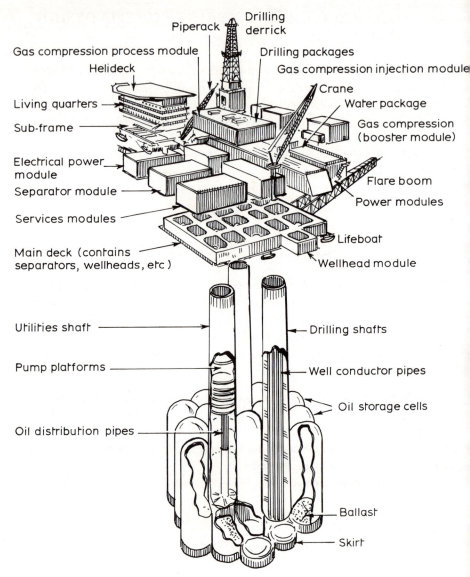

Fig. 4.3   Brent 'B' production platform[1].

seabed around the circumference of the storage tanks, and the placement of anti-scour material around the perimeter.

4.7.2   Howard Doris design

This design consists of a single main leg with an outer breakwater wall perforated with Jarlan holes. This breaks up waves, thus reducing their force.

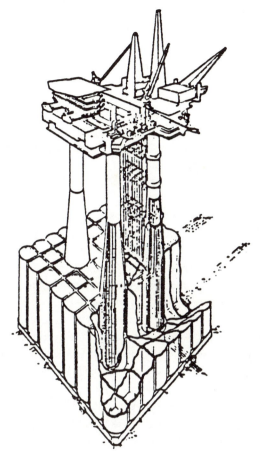

Fig. 4.4   Cormorant 'A' platform[1].

On certain designs, the pipelines are pulled into tubular tunnels which are sealed and the water pumped out, allowing pipeline tie-in to be carried out in the dry.

4.7.3   McAlpine Sea Tank design

This is a very similar design to the Condeep structure, but incorporates four legs instead of three.

4.7.4   Andoc design

This is very similar to the Sea Tank design, but the four concrete legs terminate and steel legs take over to support the deck structure.

4.8   PIPELINES

Offshore pipelines are used to transport oil or gas from the platform to loading towers or to shore. They are fabricated from high-grade steel pipe (e.g. API-5LX), anti-corrosion wrap and a weight coating. The sizes normally vary from 50 mm (2 in) to 914 mm (36 in) and the wall thickness normally varies according to the pressure rating required.

By the mid-1950s, the developments of the exploitation of oil and gas reserves offshore had advanced to the point where pipelines were required to extend over distances in excess of 80 km (50 miles) offshore. The traditional methods of pipelaying were no longer adequate and the method of using lay barges was developed.

The basic strategy was to weld lengths or 'joints' of pipe together on the barge and lay the pipe over the back of the barge by pulling the barge forward by means of a spread of anchors, the pipelines thus being lowered progressively down the side ramp and stern-mounted 'stringer' to the seabed. The barge is supplied with 12-m (40-ft) joints of concrete-coated pipe. These are welded together as a continuous process. The welded joint is X-rayed and the uncovered part of the joined pipe coated with a bitumen coating.

An alternative approach is laying pipe from a reel barge. The earliest application of this technique occurred during World War II, when a 76-mm (3-in) diameter pipe was laid across the English Channel in Operation Pluto, by unwrapping the pipe from large floating reels. The basic idea is that the pipe is prepared on land and wound onto a reel. The reel is mounted on a vessel and taken to the site and laid by unreeling the pipe by steaming the ship forward. The work on land is carried out in a spooling yard, where the pipes are supplied in lengths of about 12 m (40 ft). These are then welded together to form stalks, usually about 518 m (1700 ft) long. All the welds are X-rayed and coated, and the stalks are stowed in racks near the spooling dock.

At the start of spooling, the first stalk is moved into the roller system. The end is welded to a stub of pipe on the reel and is pulled onto the reel. The second length is then welded to the end of the first, the weld is X-rayed and coated, and the procedure is then repeated for subsequent stalks. All welding and loading operations are performed at the shore facility and so are not particularly affected by weather conditions.

Establishing and maintaining 'even tightness' of the wraps on the reel is critical in order to avoid potential breakthrough of one wrap into another, thus causing damage to the pipe. The reeling and unreeling of the pipe actually causes yielding of the steel.

4.9   OFFSHORE OIL TERMINALS

Large oil tankers are cheaper to run than small tankers. This philosophy of building large tankers was further reinforced when the Suez crisis forced tankers to detour around Africa in order to reach Europe. As tanker sizes increased, the number of ports that could handle tankers decreased and public opinion was against allowing such tankers too close to inhabited areas.

In order to meet this requirement, many solutions were proposed including artificial harbours, artificial offshore islands, multiple buoy mooring systems, tower mooring systems and single point mooring systems. The single point mooring is the most widely used because of its relatively low operational cost, reliability and flexibility.

4.9.1   Single point mooring systems (SBM)

An SBM (single buoy mooring) is basically a round buoy fixed to the seabed by chains. On the top of the buoy is a turntable mounted on large roller bearings. The tanker moors onto the buoy and rotates by any force of currents or winds into the path of least resistance. The connecting pipeline then floats out from the buoy and is connected to the tanker. At the centre of the turntable there is an oil swivel which allows for 360° rotation without impeding the oil flow.

4.9.2   Single anchor leg mooring systems (SALM)

The SALM is a buoy anchored to the seabed by chains, for example, the Mobil Beryl single point mooring. The fluid swivel is now situated at the seabed. Sometimes, deep water SALMs have the swivel assembly within the floating buoy body, and so make use of a rigid tubular riser.

4.9.3   Single buoy storage systems (SBS)

The SBS is a buoy permanently connected to a storage unit by a hinged arm.

4.9.4   Exposed location single buoy mooring system (ELSBM)

The ELSBM is an SBM specially modified for use in the severe conditions of the North Sea. The design used by Shell (UK) Ltd consists of two concentric cylinders, one on top of the other. The bottom cylinder is submerged and the top cylinder, which emerges from the water, is of smaller diameter to reduce the wave loadings. On the top of this cylinder is mounted the main roller bearing, helicopter deck, power supply, winches for handling the hawsers and hoses, and living quarters for three men.

The first parts to wear out on an SBM are the hoses and hawsers, so after use these are winched back on board the buoy.

Minimum heave motion is obtained by reducing the waterline area. This has the effect of moving the metacentre and the centre of buoyancy closer together. To increase stability, solid ballast is loaded into the bottom of the lower cylinder.

4.9.5   Components of the SBM

(a)   *Buoy body*

The SBM hull is divided into four or more watertight compartments. The design is such that should one compartment become flooded, then there is still sufficient buoyancy to support chain, hoses and piping.

A centre trunkway is open to the sea to allow the subsea pipes to be attached to the flange on the swivel. A skirt is fitted around the lower part of the hull, which houses the main fendering arrangements and anchor chains. Pipe fenders are also welded onto the outer circumference of the top of the hull. Generally, corrosion protection is by sacrificial anodes.

(b)   *Fluid swivel*

The fluid swivel assembly forms the connection between the fixed and rotating parts of the single point mooring system. The complexity of the design will depend on the number of different fluids to be handled.

It is important that no mooring forces are transmitted to the swivel head, as this could result in damage to the seals, loss of rotational motion and possible leakage.

(c)   *Mooring lines*

These consist of nylon ropes permanently attached to the buoy, and at the ship end, chains are used to avoid chafing damage to the nylon.

(d)   *Chain anchor systems*

SBM terminals use a chain system to anchor the buoy body. The number of chains used will vary from three to about 12.

4.9.6   Selection of site for single point mooring system

The main requirements for the site of any terminal are ease of access and manoeuvring capability. The depth of water must be sufficient for the largest tanker predicted to manoeuvre safely.

4.10   FUTURE DEVELOPMENTS

It would seem that with the slowing down of the development of existing fields due to government intervention in the form of higher taxes and the slump in world oil prices, both of which factors are coupled with new techniques which will increase the amount of recoverable oil coming out of existing deposits, future investment may be curtailed. If this prediction is accurate, this situation will lead to a slowing down in the development of new fields, which in turn will mean that new designs will be used for future production facilities. These same designs could also be used on existing fields of marginal economic value.

Production platforms utilizing subsea wells on templates and floating production units appear as strong contenders in this category. More conventional structures, possibly over a predrilled subsea template, may well appear in larger fields. At present, the maximum depth for a steel structure is 259 m (850 ft) (Hondo Field). This depth may be exceeded in the future, again with more unconventional designs, such as the tension leg platform design by Conoco for the north-west Hutton Field, which will possibly be the first of its type to be installed.

REFERENCE

1. Shell (UK) Ltd (1977) *The Brent System*, Shell (UK) Ltd, London.

5 Loading of offshore structures: engineering concepts

The force exerted by wind, water, weight of equipment and working loads on a structure is supported by the material of which it is built. These forces set up stresses within the material. Stress is a convenient way of defining the load we require a material to withstand in such a way that we can compare the loading on structures of different sizes and shapes. It also allows comparison with the mechanical properties; for example, how near the working stress of the member is to the yield stress or ultimate tensile stress of the material. Comparison of stresses in different parts of the structures shows those members which are carrying the heaviest load.

Stress is defined as the load divided by the area carrying that load; i.e.:

$$\text{Stress} = \frac{\text{load}}{\text{area}}$$

Stress is a loading intensity, and is a force per unit area. This idea for defining the load is widely used in fluids (including gases), and is referred to as a pressure, which is again the load divided by the area.

5.1 TYPES OF STRESS

When a material is required to support or transmit a load, it does so by creating a force between the atoms of the material by moving them from their equilibrium position.

5.1.1   Tensile stress

This is created in the material when the atoms are pulled apart (see Fig. 5.1).

$$\text{Tensile stress } \sigma = \frac{\text{load}}{\text{area}} = \frac{w}{a \times b}$$

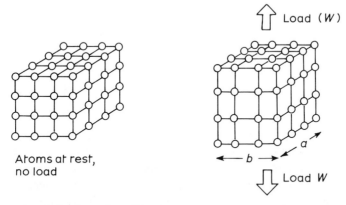

Fig. 5.1   Tensile loading of a solid.

5.1.2   **Compressive stress**

This is the exact opposite of tensile stress. The atoms are pressed together as in Fig. 5.2.

$$\text{Compressive stress } \sigma = \frac{\text{load}}{\text{area}} = \frac{w}{a \times b}$$

The symbol generally used for tensile and compressive stress is the Greek letter sigma σ. Sometimes tensile stresses are thought of as positive ($+$) stresses, and compressive stresses as negative ($-$) stresses.

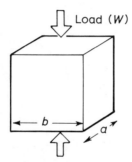

Fig. 5.2   Compressive loading of a solid.

5.1.3   Bending stresses

Sometimes a structure is loaded in such a way that there is a mixture of tensile and compressive stresses in it. A simple baulk of timber supported at the ends and loaded in the middle is a good example of this. Figure 5.3 shows the shape when load is applied.

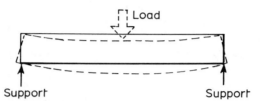

Fig. 5.3   Bent shape of a loaded baulk of timber.

The top surface is observed to get shorter as it experiences compressive stresses, and the bottom surface gets longer as it experiences tensile stresses. This type of loading gives a stress distribution that varies from maximum compressive stress on one side, to zero at an unstressed layer called the neutral axis, to maximum tensile stress at the other side. In this type of structure, we have both tensile and compressive stresses. Most braces in platform structures experience this mixture of stresses.

5.1.4   Shear stresses

The other way in which the atoms can be moved to create a force is when layers of atoms are pushed past each other. This is called shear.

$$\text{Shear stress } \tau = \frac{\text{load}}{\text{area}} = \frac{w}{a \times b}$$

The symbol generally used for shear stress is the Greek letter tau τ.

In general, fluids and gases at rest cannot produce a shear resistance when stationary, and so are subjected to pressure only, which acts at right angles to any surface.

As well as the shearing action shown in Fig. 5.4, most rotating motion is

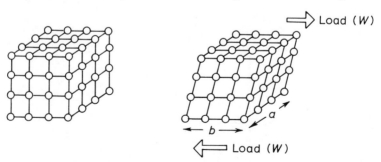

Fig. 5.4   Shear loading of a solid.

transmitted by shear; for example, the drive shaft of a car or the force to tighten a valve.

Ultrasonic waves discussed later in the book are transmitted as compressive stress waves or shear stress waves, and the atomic movement in these waves is as described above.

5.2   PROPERTIES OF MATERIALS

5.2.1   Yield stress

When a component is loaded, the material initially behaves elastically. This means that when the load is removed, the component returns to its original size and shape. This will continue while the component is in use, unless the yield load is exceeded. Yield stress is therefore the stress at which the material will no longer behave wholly elastically.

If the loading is continued beyond the yield point, the material will deform and some of that deformation will be permanent. Therefore, if a structure or part of it is dented or bent, then it has been loaded above the yield stress.

5.2.2   Ultimate tensile strength (UTS)

If loading is continued well into this region, it reaches a maximum value known as the ultimate tensile strength (σ_{UTS}). Attempts to load beyond this value will result in the material failing by ductile fracture. Ductile fracture can be identified by a large amount of local deformation in the region of the fracture.

Brittle fracture or fatigue fracture is by crack propagation without noticeable local deformation in the region of the fracture. Because of this lack of visible deformation, any technique which can assist in making it easier to detect cracks needs to be considered, and this need has been one of the main driving forces behind the development of the science of non-destructive testing. For tensile test characteristics for ductile and brittle failure see Fig. 5.5.

5.3   STRESS CONCENTRATION

It is found that in a component under load where there is a change in shape just locally at that change in shape, the stress is very much higher than elsewhere in the plate. This effect is quantified by the stress concentration factor (K_t), which is the ratio of maximum stress at that change in section to the average stress. This was first developed by Professor Inglis when studying the cause of the formation of cracks from the corners of hatch covers in the decks of merchant

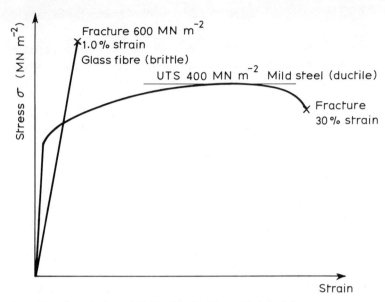

Fig. 5.5   Tensile test characteristics for ductile and brittle failure.

ships at the beginning of this century. The effect of this is best shown by considering the simple example of an elliptical hole in a flat plate. The flat plate is loaded in tension in one direction only with a stress, σ, and in the plate is an elliptical hole shown in Fig. 5.6. The plate is so large that the loss of area due to the cutout can be neglected. At the edge of the cutout at points A, the stress will be higher than the average stress σ by an amount given by the stress concentration factor K_t, such that the stress at $A = \sigma_{max} = K_t \times \sigma$. The value of K_t for this case is given by the equation:

$$K_t = 1 + \frac{2a}{b}$$

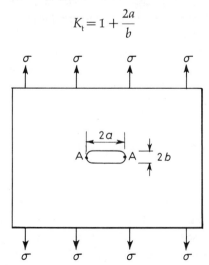

Fig. 5.6   Load on a flat plate elliptical hole.

It is worth examining this equation in a little more detail. Figure 5.7 shows the variation of K_t with the shape of the hole. It will be noticed that the shape of the cutout and its position in the stress field is important. The longer and thinner the cutout (or the defect), then the larger the stress concentration factor K_t and hence the higher the maximum stress at the tip of the defect.

The position of the cutout or defect in the stress field is important. If the long side of the cutout is at right angles to the maximum stress, then the stress concentration factor is higher than if it runs in the direction of maximum stress.

An awareness of the fact that stresses are higher at changes of shape alerts the inspector to areas most at risk from cracking by fatigue. As there is seldom sufficient time to inspect the whole of a structure, this knowledge leads to a more effective inspection programme. Some of the changes in shape that will give rise to stress concentrations and hence lead to increase in stresses in the structure are as follows:

(a) Node joints.
(b) Cutouts in members.
(c) Holes in drilling templates.
(d) Weld profiles (unless ground flush with the parent metal).
(e) Weld defects.
(f) Corrosion pits.
(g) Marks on the surface of the material caused by accidental damage or bad workmanship.

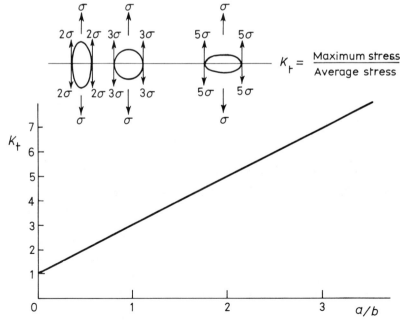

Fig. 5.7   Variation of stress concentration factor (K_t) with shape.

As a crack grows, the value of K_t will increase. Because the value a/b increases, this is sometimes referred to as sharpening the crack.

One of the standard emergency remedial procedures when a crack has been discovered, is to drill a hole at the tip of the crack, often referred to as crack blunting or stopping. The reason for this can be understood because the stress at the tip of the crack will be high. Further growth of the crack will increase the value of K_t to an even higher value (for a long thin ellipse, this can be greater than 10). The drilling of the hole at the tip of the crack reduces the stress concentration factor to three and hence reduces the high local stress at the tip of the crack, making the crack less likely to propagate; i.e. grinding out cracks reduces the a/b ratio.

5.4   RESIDUAL STRESSES

Residual stresses are stresses that have been set up in the structure during manufacture. They may have arisen from thermal stresses caused by welding a fabricated structure, or by mechanical stresses set up by force fitting members of a structure. These can be removed by a stress relieving treatment. If, however, residual stresses remain in a material, the working stresses set up will add to the value of the residual stress. This means that the structure is subjected to a higher stress in service than the design predicted, as the designer would have calculated only the working stress.

5.5   FORCES ON A STRUCTURE

The forces that the structure experiences are of two types: steady and vibrational. These forces are produced by several different effects; for example, the weight of the equipment, the reaction of the drilling force, and the force exerted by the wind and the water.

5.5.1   The steady force on a structure in a fluid flow

The steady force exerted by a fluid as it passes a stationary structure is known as the drag force. Therefore, if a structure is placed in a current of water (the tide) or air (the wind), it will experience a force in the direction of the flow trying to move it in that direction. This can be illustrated by placing a walking stick in a swiftly flowing stream. A holding force must be exerted to keep the stick in position. This holding force is equal and opposite to the drag force on the walking stick caused by the stream (see Fig. 5.8).

The size of this drag force depends on several factors which are related by a simple formula:

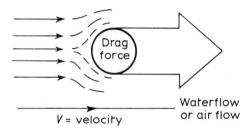

Fig. 5.8   Plan view of fluid flow past a cylinder.

$$\text{Drag force} = \frac{1}{2} C_d \rho A V^2$$

In this formula V is the velocity of the fluid flow. Note that the force on the cylinder in the flow varies with the square of the velocity. For example, double the flow speed and the drag force is increased by four times; treble the flow speed and the drag force is increased by nine times.

A is the projected area at right angles to the direction of the fluid flow, which for a cylinder of diameter D immersed to a depth h is given by $A = D \times h$.

ρ is the density of the fluid. As water is denser than air, the drag force on a cylinder in water would be greater than that of air flowing at the same velocity.

C_d is the drag coefficient. This is a number that takes into account the shape of the structure and the roughness of its surface. For example (see Fig. 5.9), consider a disc of diameter D in a fluid flow of velocity V. If a hemisphere of the same diameter as the disc is fixed on the front of the disc, we find that the drag force is reduced. As nothing else has changed, it means that the drag coefficient of this shape is less than that of the disc. If a cone is now added behind the disc, we find that the drag force is further reduced. Again, as nothing has changed, it means that the drag coefficient has reduced. This process is of course known as streamlining and it is considered for any vehicle that moves through a fluid (e.g. boat, aeroplane, submersible) and any structure that is in a moving fluid stream

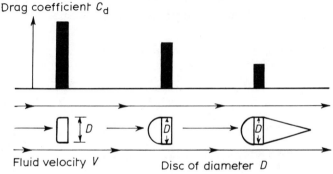

Fig. 5.9   Change of drag coefficient with shape.

(e.g. platform, jetty piles, pylons and aerials), provided that there is a constant known direction of flow. If not, cylindrical members provide the best compromise.

5.5.2   Vibrational forces on a structure in a fluid flow

If we return to our previous example of a cylinder in a fluid flow and this time look at the flow pattern behind the cylinder, we notice that it is not symmetrical, but that vortices are shed alternatively from each side (see Fig. 5.10).

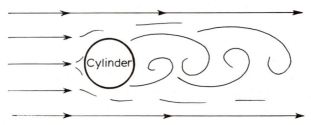

Fig. 5.10   Von Karman vortices shed from alternate sides of a cylinder in a fluid flow.

The effect of this is to place on the cylinder an alternating force at right angles to the fluid flow and drag force direction (see Fig. 5.11).

These cyclic forces generated by the wind and water flowing past the structure cause the vibrations that are so important when considering the fatigue life of the structure.

5.5.3   Wave loadings

Waves provide an oscillatory motion to the structure, producing forces which act in addition to the forces produced by tidal currents. These forces deform or try to overturn the structure. The waves have a predominant direction for their maximum effect, but can come from any direction, since they are wind generated. Waves produced in a storm are generally short and very confused.

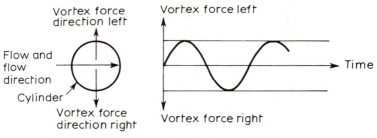

Fig. 5.11   Variation in side forces on a cylinder in a fluid flow.

However, when produced by winds blowing over a long distance or fetch, the waves tend to moderate into long, high swell waves with a long period. A period of 14 seconds produces a wavelength of about 300 m. The height of the waves is independent of the period, but depends upon the stability of the waves and the energy content.

For the purposes of classification and insurance, there are standards for any design, and these are based upon statistical data. As much information as possible is collected over as long a period as possible, on wave heights, directions and periods. The analysis of these data produces two main results:

(a) The maximum wave to be expected in a given time span, generally a 100-year period. (This is only a statistical quantity adopted for design, so more than one of these waves, or even larger waves, might actually occur.)
(b) An energy spectrum of the waves (i.e. the graph of the energy in the waves at different periodic times).

Structures are then designed for two conditions:

(a) Static loading, using the maximum 100-year storm wave.
(b) Dynamic loading, using the energy spectrum.

Owing to the directional properties of the waves, the structure will be designed and placed so that the largest waves from the predominant direction are taken on its strongest orientation, but all other directions should be considered. Inaccuracy in placing the structure can create loads greater than the design loads in that direction.

The static analysis based on the 100-year storm wave is the simplest to perform, but must result in a great deal of work, since both the direction and position of the wave crest, relative to the structure, will produce different effects on different parts of the structure.

A wave, being an oscillatory motion, contains water particles with both velocity and acceleration. The velocity will produce drag forces, as mentioned above, and the acceleration will produce inertia forces, in the same way as any car which is slowed down requires a braking force.

Considering the wave force on a circular member (diameter D) submerged to a depth (L) of a jacket:

$$\text{Total wave force} = \text{drag force} + \text{inertia}$$

$$= \frac{C_d}{2}\,\rho D L V^2 + (\text{mass} \times \text{acceleration})$$

where C_d is the drag coefficient, ρ is the density of water and V is the velocity of the water.

The mass considered is the mass of water which moves around the structure, which equals:

$$\text{Coefficient} \times \rho \times \frac{\pi D^2}{4} \times L$$

The coefficient is found by experiment to have a value between 1.5 and 2.0, and is called the inertia coefficient (C_m), but this will depend upon the roughness of the surface and the period and height of the wave.

The wave force is equal to:

$$\left(\frac{C_d}{2} \rho D V^2 + C_m \rho \frac{\pi D^2}{4} \times a \right) \times L$$

where a is the acceleration of the water.

Note the importance of the diameter D in these expressions, and hence the necessity of keeping the thickness of marine growth to a minimum. The direction of this force varies with the position of the wave. The above expression applies only to members which have a small diameter relative to wavelength, and this applies to most steel structures. The forces on large concrete gravity structures are found by considering the pressure variation around them.

The dynamic response of a structure can be demonstrated easily by swinging a weight on the end of a string, as a pendulum. If the movement of the hand holding the string varies in frequencies, so will the deflection of the weight at the other end of the string. If the amplitude of the hand movement is fixed and the swing or displacement of the weight is noted, it will be observed that as the frequency of oscillation of the hand increases, so the amplitude of the swinging weight will change. First, the amplitude will increase up to a maximum value; the frequency of the hand movement at this condition will be the natural frequency of the system, and thereafter the amplitude will decrease.

Another observation is that as the weight displacement increases, the direction of the swing will be the same as that of the hand movement. The motion of the hand and the weight are in phase. Once the maximum amplitude of the weight has been exceeded, it will be noticed that the weight is moving in the opposite direction to the hand movement, and the motion is then said to be out of phase. As the frequency of the hand motion is further increased, the amplitude of the weight decreases to almost nothing.

The peak displacement of the weight occurs at a point called the natural frequency. This will decrease with increasing length of string and increasing weight. Thus, at certain frequencies of forcing the hand movement of the example, the displacements produced can be very large.

The same process occurs with an offshore structure. A structure will be of a certain height, construction and weight, and these will determine its natural frequency. The forcing frequency will be supplied by the waves, or rather by the waves' energy spectrum.

When a structure is placed in the sea, it will experience a range of wave energies and frequencies, causing the structure to deflect. As the frequency of the wave energy peak approaches the natural frequency of the structure, so the deflection of the structure increases, and with it the stress. The further the peaks of wave energy, frequency spectrum and natural frequency are separated, the lower the maximum deflection of the structure.

This same analysis applies to diving and other floating vessels in heave, roll and pitch. Thus, vessels designed for use in one part of the world may be unsuitable for use in another, where the wave energy spectrum differs.

The natural frequency decreases as the height of the structure increases. Thus, new designs of structures are being developed for deep water applications, such as the guyed tower and the TLP (tethered leg platform), which have natural frequencies below the wave energy peak, as do most floating vessels. It can be envisaged that the deflections produced at some frequencies could be much greater than that produced by the static application of the 100-year storm wave, and so the stresses would be correspondingly higher at these conditions.

It is evident that the small waves which occur at frequencies around the natural frequency of the structure produce very high deflections, and since they exist in large numbers during the life of the structure, they are likely to be more damaging than the one or two large waves which may occur during that lifetime. This explains many of the problems now being experienced in the structures, since only the very last structures installed were designed with a view to the dynamic effects being significant.

The oscillatory nature of the deflections of the structure caused by the waves produces a large number of stress reversals in the structure, in the same way as a diver experiences forward and backward motion in the surface zone. These stress reversals provide the loading that produces the fatigue damage which most structures experience.

5.6   MECHANICAL DETERIORATION IN SERVICE

As soon as a piece of engineering equipment, such as an engine, pipeline, bridge or offshore structure, is brought into service, it starts to wear out because of use, and if it is not maintained it will eventually cease to operate satisfactorily, either by no longer carrying out the function for which it was designed or by failing in a catastrophic manner. The possible causes of deterioration of an offshore structure are discussed below and include accidental damage, corrosion, fatigue, wear and embrittlement.

5.6.1   Accidental damage

As the structure is served by boats, there is a real possibility that the structure could be damaged by accidents such as collisions and the dragging of anchors across seabed installations. Ideally, these should be reported and surveyed as soon as they occur, but often such accidental damage is only noticed during a diver's routine inspection.

5.6.2   Corrosion

Because steel is placed in a hostile environment, namely salt water, one of the ever-present deterioration mechanisms on the structure will be corrosion.

Corrosion takes place in two different ways. First of all, uniform corrosion is the process whereby metal is removed uniformly from all over the surface, so that progressive thinning of the member or pipe wall goes on until the thickness is reduced so as to necessitate the renewal of the component.

Secondly, pitting corrosion is a very localized corrosion which takes place in an otherwise corrosion-free material, creating a pit in the surface of the material. These pits deepen with time and if another failure mechanism does not take over, the pit will penetrate the full thickness of the material, causing leakage in the case of a pipeline or service duct, and so necessitate local repair.

Corrosion attacks of both kinds are accelerated by erosion, increase in temperature, increase in oxygen content, added chemical attack from biological sources and loading on the member from either external loading or residual stresses caused in manufacture. This is known as stress corrosion.

5.6.3   Fatigue

Fatigue is the local failure of the material by crack growth caused by cyclic loading. The cracks can grow from flaws in the material, such as a welding defect or notch caused by accidental damage. Alternatively, they can initiate in regions of highly stressed material which are brought about by residual stresses or stress concentration. Fatigue cracks can also start from pits created by corrosion. This condition is known as corrosion fatigue.

5.6.4   Wear

This is the thinning of material due to uniform corrosion or erosion or a combination of the two.

5.6.5   Embrittlement

In this case, the material changes its properties from being ductile to brittle. This can be a localized effect. Brittle materials fail due to crack propagation so that they are susceptible to fatigue as well as to brittle fracture. Embrittlement in service could come about due to incorrect welding procedures or by the absorption of a gas, generally hydrogen. Embrittlement has been encountered in natural gas pipelines and could come about from the absorption of hydrogen produced in an overprotected corrosion system.

5.7   FAILURE OF THE STRUCTURE

This will generally occur in one or a combination of modes. The following are probably the most significant, but with failure sometimes an unsuspected mechanism will cause the structure to become unserviceable.

(a) Fatigue cracking followed by collapse.
(b) Brittle fracture originating from sites such as fabrication defects, fatigue cracks and/or local accidental damage.
(c) Failure resulting from corrosive attack.
(d) Collapse due to extreme static or dynamic loads.

6 Deterioration of offshore structures

With every engineering structure or even in a component of a structure it must be realized that the danger exists of including defects into it at any stage of manufacture or fabrication. These defects could then lead to component or structure failure at any time after final fabrication or during or after installation. Of course the quality control of the manufacturer will obviate this problem and the tighter the design specifications and fabrication procedures and the greater the degree of quality control that is exercised the better will be the finished structure. It must be appreciated, however, that structural deterioration begins with the start of the fabrication of the structure and continues throughout its life.

From this point of view the life of an offshore structure can be divided into four stages, and defects can be created at any stage. Looking at steel structures at this point then the stages are as follows.

6.1 STAGE ONE: PRODUCTION OF THE STEEL

During production of the raw material, several defects can be included into what will become the parent plate. Examples of these problems are as follows.

6.1.1 Fish tails

These may occur in steels produced by traditional mills. In these mills the process proceeds in stages. The first is the smelting of the steel and this is followed by pouring of the molten metal into a crucible, which is used to transport and then transfer the steel to the ingot moulds. The transfer into the moulds is done by pouring and at this point some of the liquid steel can splash up the sides of the mould. This will then cool on coming in contact with the cold sides of the mould, and will solidify. The rest of the liquid steel continues to be poured in and then covers the solidified splashes. When the ingot is removed from the mould, the splashes, which form shapes similar to fish tails (hence the name), adhere to the surface. If these fish tails are not removed before rolling they may be rolled into the surface without bonding, thus causing a reduction of material thickness. Figure 6.1 illustrates this.

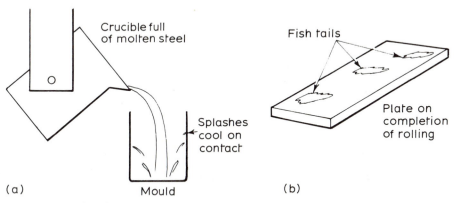

Fig. 6.1   Fish tails.

6.1.2   Laminations

Again in traditional mills, the process may be responsible for causing a plate defect called lamination. When the molten steel is allowed to cool in the mould it shrinks. This causes a reduction in volume which manifests itself as a pipe. When the ingot is removed from the mould and proceeds through the rollers it becomes very thin, in fact cracklike. This is the lamination as illustrated in Fig. 6.2.

6.1.3   Lamellar tears

With plate which is thicker than about 40 mm (1.6 in) there is the possibility that micro fissures can be formed along the grain boundaries. This is because the pressure is not high enough to forge these fissures together as the plate cools. If the plate is then stressed in the through plate direction during service, these fissures may grow and form lamellar tears.

6.2   STAGE TWO: FABRICATION

During this phase the main problems are associated with welding. During the welding process (as with the manufacturing of the steel plate) various defects

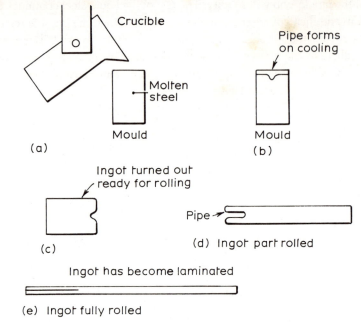

Fig. 6.2   Laminations.

can be included into the weld which may cause failure later. Some examples of these defects will illustrate this.

6.2.1   Lack of root penetration

This can be a fabrication defect which occurs with both submerged arc welding and with manual metal arc welding. Both of these processes are used or have been used extensively on offshore platforms. The defect is caused in both cases by the operator or welder. In the case of the submerged arc process the incorrect voltage is set and with the manual metal arc the welder does not achieve the correct position with the welding rod.

6.2.2   Slag inclusions

These are possible at any time when the manual metal arc is the method being employed with a multi-pass technique. Once again, this technique is or has been used on offshore structures. In this case the defect is caused by the slag being incompletely cleaned out between the weld runs.

6.2.3   Porosity

There is always a potential danger of porosity occurring with manual metal arc welding. The source of the gas that causes porosity is usually air, dirt and damp, which produce nitrogen, hydrogen or carbon monoxide, which dissolve in the liquid metal weld pool. When the weld pool cools, the gas comes out of solution and forms bubbles of gas in the weld metal.

6.2.4   Hydrogen-induced cold cracking

This is cracking associated with the heat-affected zones and occasionally with weld metals of ferritic steels. It can occur immediately at temperatures as low as $150°C$, or sometimes only appears hours after welding. It usually originates at the toe of the weld. The hydrogen dissolved in the weld pool diffuses into the heat-affected zone of the parent metal in sufficient quantities to embrittle the martensitic structure and cause cracking. Even if cracking does not occur on cooling, the dissolved hydrogen in the steel makes it more susceptible to crack propagation under in-service loading.

6.3   STAGE THREE: INSTALLATION

During this phase of the life of a structure, the structure itself may be subjected to much greater stresses than at any other time of its existence. This is because of the method of installation of these steel structures. The structure itself is built on its side and is floated and towed to its destination in this attitude. Once in location, the structure needs to be tipped upright and although this is done in the most gentle way possible using the surrounding water as a cushion, the moment of righting the structure to the vertical position still imposes considerable stress. This is shown in the case of the self-floating type of structure, below.

6.3.1   Self-floating structures

These structures have flotation units to enable the structure to be towed to site. On site, the units are flooded in a pre-set sequence to right the platform and lower it to the seabed. The flotation may be provided by large diameter legs (up to 9 m (30 ft) in diameter) split into compartments; for example, Brent A (see Fig. 6.3). These installation stresses may in some cases lead to the structure being designed to withstand the installation stresses rather than the operational loads, which may be lower. Finally, there exists the possibility of parts of the structure being overstressed due to unforeseen loadings, which may cause cracking at this stage.

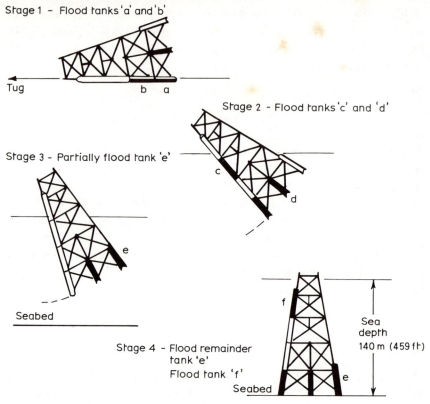

Stage 1 – Flood tanks 'a' and 'b'

Tug   b   a

Stage 2 - Flood tanks 'c' and 'd'

Stage 3 - Partially flood tank 'e'

c

d

e

Seabed

Stage 4 - Flood remainder
             tank 'e'
          Flood tank 'f'
                  Seabed

f

Sea
depth
140 m (459 ft)

e

Fig. 6.3   Placing Brent 'A' on the seabed.

6.4   STAGE FOUR: IN SERVICE

During this stage, the structure is subjected to:

(a) All the stress imposed on it by the elements.
(b) A build-up in marine growth increasing structural loading.
(c) Deterioration caused by corrosion and possibly by additional loading being imposed on the structure by the addition of extra plant or modules to the platform.
(d) Accidental damage to the structure caused either by the elements or by mishap.

Accidental damage will manifest itself in the following way.

6.4.1   Loss of member straightness

This can be caused by damage at the installation stage, as was the case with the north-west Hutton Field, where an accident at that time led to loss of member straightness on the main leg. An alternative cause of this type of damage can be accidental impact caused either by equipment being dropped overboard or by boats coming into contact with the structure.

6.4.2   Local deformation of individual members

Once more, accidents at the installation stage may be the cause of this type of damage. For example, a large indentation was caused by a pile which was dropped during the piling operation on one North Sea platform. This damage necessitated a repair being undertaken even before the structure was put into service. In service, all manner of equipment is at risk of being dropped overboard and can thus cause this type of damage.

6.4.3   Weld failure

It is possible that latent defects caused at the fabrication stage may lead to failure in service when they become stressed. Alternatively, fatigue caused by cyclic loading during service may also be responsible for weld failure.

6.4.4   Progressive overloading

The build-up of marine growth on a structure has the effect of increasing the static load on the structure because of its weight. It may also increase the dynamic loading on the structure, because it will increase the overall drag on the structure. The effects of marine growth on a structure are more noticeable because of the fact that the majority of the growth will occur in the top 30 m (98 ft) of the water and thus the weight and drag are concentrated in this area.

During the working life of the structure it is not uncommon for additional equipment and even additional modules to be added. This will also add to the loading on the structure.

Drill mud will collect around the structure and this can itself cause overloading because of the additional weight this imposes on the structure.

Finally, debris of all kinds will collect around the structure, some falling all the way to the seabed and some fouling up within the fabric of the structure itself. Once again this will cause overloading of the structure.

6.4.5   Deterioration caused by corrosion

The amount of excessive deterioration caused by corrosion will depend on the actual corrosion rate. The rate of corrosion will be affected by:

(a) The amount and type of marine growth.
(b) The amount and type of debris.

(c) The type of corrosion, whether it is pitting or uniform corrosion.

(d) Whether or not there is localized corrosion.

(e) In the case of impressed current systems, any loss of electrical continuity for whatever reason will lead to loss of protection in the local area.

Taking these factors into account the effects of excessive corrosion will be a reduction in wall thickness which in effect will cause a local overloading of the structure.

6.4.6   Accidental damage

There is no point in speculating as to precisely what causes accidental damage, because by its very nature it is unpredictable. What can be said with some confidence is that this type of damage is largely not reported. This in turn can mean that quite significant damage can be caused to the structure without its being recorded. It is not possible to predict this damage and it is therefore not possible to include this into an inspection programme.

6.5   SEABED INSTALLATIONS AND PIPELINES

Seabed equipment of all types will be subjected to similar problems with regard to wear and damage and comments made earlier about structural deterioration also apply to these installations. Because of their situation, however, they suffer from certain problems peculiar to themselves. These particular points follow below.

6.5.1   Debris

All types of debris from tools to scaffold bars can and do fall off structures and vessels in and around offshore oilfields. This debris can all land on seabed installations, causing any amount of damage.

Similarly, drill mud can collect in large quantities on the seabed and may bury part or all of the installation. Vessels anchoring in the area of seabed installations may drag their anchors and this can lead to damage.

6.5.2   Damage to pipelines

Pipelines can be affected in a similar manner and may also suffer from damage caused by trawlers, which fish in some of the same areas as the oilfields. An additional problem with pipelines can be caused by scouring the seabed away from the pipe which leaves spans along the pipe. Short spans may be tolerated but longer areas of spanning cannot be tolerated because the pipe may well sag under its own weight, causing excessive stressing in that area.

This problem may be more troublesome in areas where rock outcrops occur because the effects of scour in this situation will be much more severe. It is also possible for rocks to form pinnacles under the pipe which will increase the loading on the pipe in that area.

6.5.3   Further effects of scour

Scour is not restricted to pipelines by any means. Its effects can be seen all around the legs of offshore platforms where the profile of the seabed is changed as the soft material is eroded away. Figure 6.4 outlines the effects of scour in this situation.

Quite apart from the effects of scour, the fact that the currents around seabed installations, pipelines and structures pick up and transport particles leads to erosion. The current-carried particles act as an abrasive to wear away the surfaces with which they come into contact. In the case of a platform, this can mean that the legs are being worn away quite rapidly right on the mud line. This loss of wall thickness in this particular spot could lead to a dangerous increase in stress.

6.6   MARINE GROWTH

The build-up of marine growth will have two effects. First it will increase the area of the profile presented to the water flow and so increase the force on the structure. Secondly, marine growth will change the texture of the surface from a smooth, round steel or painted surface to a surface made much rougher by the presence of marine growth on it. This roughness will increase with time as the surface becomes more irregular due to parts of the dead marine growth sloughing off. The effect of this is to increase the drag coefficient. Both these effects increase the force on the structure. Information on the types and amounts of marine growth build-up is required to confirm or modify the design-predicted loads on the structure. See Fig. 6.5.

These two effects of marine growth will have a knock-on effect with the structure which will manifest itself in the following manner.

(a) By producing an increase in mass without any significant change in stiffness. This causes a reduction in its natural frequency.
(b) By increasing the added mass of water and the drag forces on the structure. Marine growth being most abundant at and just below the water level coincides with the zone of maximum wave and water force, so that the forces on the structure are increased in the region of maximum water force.

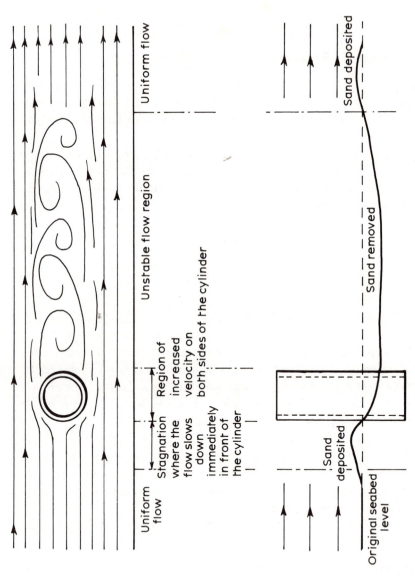

Fig. 6.4 Flow conditions and effects around a cylinder: (a) bunching of lines indicates an increase in velocity; (b) profile of a sandy bottom around the base of a cylinder.

Uniform flow

Unstable flow region

Uniform flow

Stagnation where the flow slows down immediately in front of the cylinder

Region of increased velocity on both sides of the cylinder

Uniform flow

Sand deposited

Sand deposited

Sand removed

Sand deposited

Original seabed level

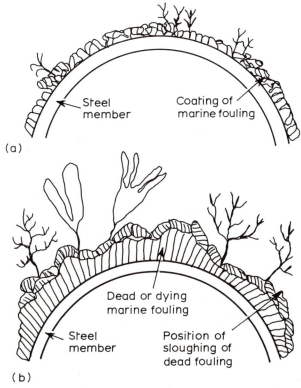

(a)

(b)

Fig. 6.5 Changes in the surface of the structure due to marine growth: (a) steel with a layer of marine growth; increase in drag force because of increased diameter and further increase due to change in roughness; (b) steel with thick layer of marine growth, some of which has sloughed off; increase in drag coefficient due to change in roughness and shape as dead fouling has sloughed off.

(c) By affecting the corrosion rate, either by accelerating or retarding it.
(d) By reducing the effective area of the service inlets and outlets, hence reducing system efficiency.
(e) By obscuring the important features on the structure, such as diver orientation marks, valve handles and handrails.
(f) By making inspection impossible before cleaning.

Marine growth is of such importance because of these effects that it is as well to examine this problem in a little more depth.

6.6.1   Types of marine growth

From an engineering point of view, there are two main categories of fouling: soft fouling and hard fouling. Soft fouling is caused by those organisms which have a density approximately the same as seawater. They are important because of their bulk, but are generally easy to remove. Organisms causing hard fouling are much denser and more firmly attached to the structure and are therefore difficult to remove.

6.6.2   Soft fouling

Organisms in this group are as follows.

(a)   *Algae*

This is often referred to as slime and is generally the first to inhabit an offshore structure. As it is very light-sensitive, it is seldom observed in any quantity below 20 m (66 ft).

(b)   *Bacteria*

This, like algae, will be amongst the first inhabitants of an offshore structure and will be present to depths well in excess of 1000 m.

(c)   *Sponges*

These are often found as a fouling species on offshore structures and are present at depths greater than 1000 m. See Fig. 6.6.

(d)   *Sea squirts*

These are soft-bodied and sometimes grow in large colonies. See Fig. 6.7.

(e)   *Hydroids*

These grow in colonies and from their appearance can be mistaken for seaweed,

Fig. 6.6   A sponge.

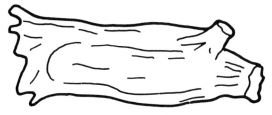

Fig. 6.7   A sea squirt.

but they are in fact animals related to the sea anemone. The colonies can produce dense coverage to depths of 1000 m. See Fig. 6.8.

(f)   Seaweeds

There are many types of seaweed that attach themselves to underwater structures, but of these, kelp produces the longest fronds, which can grow up to 6 m (20 ft) in length under favourable conditions. See Fig. 6.9.

(g)   Bryozoa

This has a moss-like appearance, grows very tall and is really an animal with tentacles. See Fig. 6.10.

(h)   Anemones

These are sometimes called anthozoans, which means 'flowering animals'. The cylindrical body is surmounted by a radial pattern of tentacles. It attaches itself

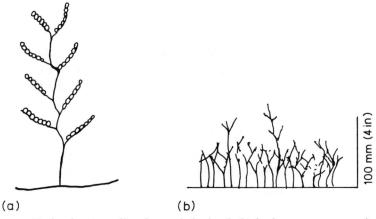

(a)   (b)

Fig. 6.8   Hydroids: (a) profile of a single hydroid; (b) feathery appearance of a colony of hydroids.

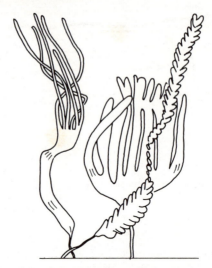

Fig. 6.9   Seaweed: profiles of three varieties of kelp weed.

Fig. 6.10   Bryozoa.

to the structure by a basal disc, and this attachment is so firm that attempts to remove it often result in tearing the body of the anemone. The colours and shapes are extremely variable even within the same species. See Fig. 6.11.

(i)   *Dead men's fingers* (Alcyonium digitalum)

Colonies have been observed on pier piles, rocks on the foreshore and offshore structures. These colonies often grow to 150 mm (6 in) in length. When submerged, many small polyps arise from the finger-shaped, fleshy main body, each polyp having eight feathery tentacles. It is white to yellow or pink to orange in colour, but when out of the water it is flesh coloured and the similarity to the human hand gives it its common name. See Fig. 6.12.

Fig. 6.11   Anemone surrounded by mussels.

6.6.3   Hard fouling

Composed of calcareous or shelled organisms, the common types in this group
are as follows.

(a)   *Barnacles*

The common species is *Balanus balanoides*. These grow in dense colonies to a
depth of 15–20 m (49–66 ft), but are observed to depths of 120 m (394 ft). See
Fig. 6.13.

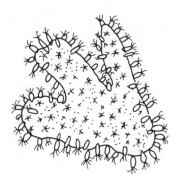

Fig. 6.12   Dead men's fingers.

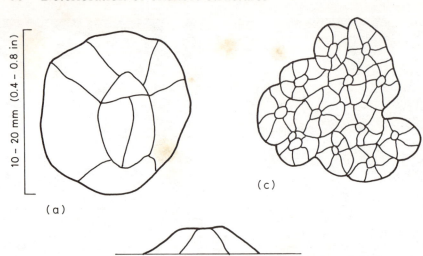

10 – 20 mm (0.4 – 0.8 in)

(a)

(b)

(c)

Fig. 6.13   Barnacle: (a) view of the top of a barnacle; (b) side view; (c) top view
of a cluster of barnacles giving a sort of hexagonal pattern.

(b)   Mussels

The main spcies is *Mytilus edulis.* This hard-shelled mollusc attaches itself to the
structure by byssal threads at the hinge of the shell. These thread attachments
are very strong and mussels generally form dense colonies. Main colonization is
to depths of 20 m (66 ft), but mussels are found to depths of about 50 m
(164 ft). See Fig. 6.14.

Fig. 6.14   Mussel colony on a bracing.

(c)   Tube worms

The full title for this species is calcareous serpulid tube worm. This often forms on flat surfaces, as shown in Fig. 6.15. It is white in colour, very firmly attached to the surface of the metal and difficult to remove. It also grows in colonies and these have been known to fill a warm water outlet, arranging themselves parallel to the flow to obtain maximum nutriments. Power cleaning is required to remove this growth, so firmly is it attached. Although the main growth occurs to depths of 50 m (164 ft), tube worms are found to depths of 100 m (328 ft).

Fig. 6.15   Tube worms on a bracing.

6.6.4   Factors affecting marine growth

If no steps are taken to prevent growth, such as application of an anti-fouling solution, then formation of bacterial slime occurs in two to three weeks. Barnacles and soft fouling have been known to attach themselves and reach maturity in three to six months. It generally takes two seasons for mussel colonies to develop, often on top of the dead earlier fouling. The type of organism, its development and growth rate will depend on several factors, including the following.

(a)   Depth

Figure 6.16 gives the generally accepted diagrammatic representation of the combined effects of weight and volume of the various types of marine fouling in British waters. The diagram shows clearly that most weight is added in the

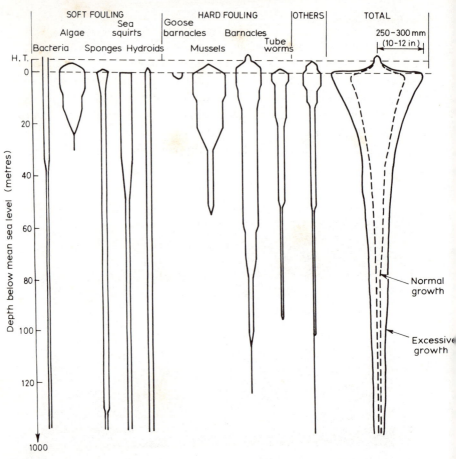

Fig. 6.16   Diagrammatic representation of the distribution of marine growth with depth.

vicinity of the surface which is the region of highest water-induced loading. The total column in the diagram is not the sum of the others, but an estimate of a balanced colony. Note that the long lengths of seaweed have not been included. Increase in depth reduces light intensity which therefore reduces the ability of organisms such as algae to photosynthesize. Algae therefore gradually disappear with depth and there is also a change in species to red algae at the greater depths. Algae growth at depths below 30 m (98 ft) has been observed in the North Sea due mainly to the clarity of the water.

(b)   Temperature

In general, a rise in water temperature will increase the growth rate of a colony. The growth rate approximately doubles with a 10°C rise in temperature. There will of course be a limit and most organisms cease growth at 30°–35°C. As the temperature variation is greatest near the surface, there is seasonal growth in the marine colonies near the surface, and continuous slower growth as the depth increases.

(c)   Water current

The speed at which the water flows over the surface plays an important part in the type of fouling colony that develops. There are two aspects to consider, the first being that of the larvae attaching themselves to the structure. It is suggested that at speeds greater than 1 knot, many larvae are unable to attach themselves. However, once attached, most fouling can withstand water currents of more than 6 knots. At high water velocities, weakly attached fouling is removed leaving only the firmly attached hard fouling. Also, colonies growing on dead or dying fouling become loose and may be sloughed off. The larvae can attach themselves to structures during slack flow periods, or in localized spots of slower flow or dead water, such as crevices and locations between hard fouling. The second aspect to consider is that in general, once the organism is established, a strong current brings more food and growth is accelerated.

(d)   Salinity

In nearly fresh water, fouling is usually confined to an algae slime. As the salinity increases, so the amount and type of fouling increases. First hydroids and barnacles and finally mussels occur. The normal salinity of seawater is about 3%–3.5% and the size of mussels, for example, increases fivefold from a salinity of 0.6% to 3.5%.

(e)   Food supply

Growth of the fouling is obviously dependent on the quantity of nutriment available. Growth rates seem to be faster in coastal waters than those a few miles offshore where the water is deeper. Recent investigations suggest that the slow currents that circulate around platforms become enriched with nutriments from sewage and other waste which will increase the growth rate.

(f)   Cathodic protection

There are two types of corrosion protection widely used on North Sea structures (see Chapter 9). On those structures that use sacrificial anodes, the patterns of marine growth on the structures themselves seem normal, but the anodes generally remain clear of growth. The other system, which uses an impressed current to cancel the corrosion-induced ionic currents between the structure and the sea, suggests, on a limited amount of preliminary evidence, that the marine growth rate is increased. At the moment, the mechanism that encourages an increased growth rate is not understood, and many more data are required before the foregoing observation is confirmed.

6.7   DETERIORATION OF OFFSHORE CONCRETE STRUCTURES

6.7.1   Concrete

This material is made up of a mixture of crushed stone or gravel (called coarse aggregates) and sand (referred to as fine aggregate) which are both mixed with cement, which is a bonding agent. Water is added to the mixture which causes a chemical reaction to take place in the cement, which then hardens to bond together all the aggregates to form a stone-like material. This material does in fact share all the basic properties of stone and, like stone, is therefore extremely good at withstanding compressive loadings. It is very poor, however, at withstanding tensile stresses.

Concrete has a tremendous advantage over stone in that it can be cast into almost any shape and can be worked as a semi-liquid paste until it sets hard. This latter property is utilized to transform concrete into a building material which has all the advantages of stone for durability and strength and unlike stone has the ability to withstand tensile stresses. This material is reinforced concrete and it is made up of a lattice of steel bars known as the 'reinforcement', over which the wet concrete mix is poured and cast into the finished shape. The durability of a concrete structure is directly proportional to the efficiency of the cover in protecting this reinforcement. The loss of the concrete coating can be caused by the environment to which the concrete is exposed or by causes within the concrete itself. Even so it is rare that one isolated case will result in deterioration; rather, it is generally the result of a culmination of factors.

There are two principal forms of deterioration of concrete: chemical and physical.

6.7.2   Chemical deterioration

It is unlikely that any significant deterioration will occur in concrete of the quality that is normally specified for offshore structures. Nevertheless, deterioration may occur in concrete which has not been properly compacted or through environmental pollution.

(a)   Sulphate attack

Cement contains a proportion of tricalcium aluminate which takes part in the hydration process. This can react with magnesium sulphate, which is present in concentrations of about 0.5% in seawater. The reaction is expansive, but the presence of chlorides inhibits the degree of expansion. The net result is softening and disruption of the concrete in the form of solution and crumbling.

(b)   Chlorides

Chlorides do not attack plain concrete when present in the concentrations that normally exist in sea water, but they may greatly accelerate the corrosion of reinforcing steel by destroying the passivity of the concrete coating.

(c)   Carbonation

Carbon dioxide is present in the air and can attack the concrete directly. It has the effect of destroying the normal passivity of the concrete coating, thus leading to reinforcement corrosion, but its effects are generally limited to a penetration of approximately 50 mm (2 in).

(d)   Contaminants

There are various substances that may be inadvertently incorporated into the concrete matrix (e.g. glass, coal, gypsum) which can react with the cement in the presence of seawater.

6.7.3   Physical deterioration

Physical deterioration can result from an expansive pressure in the concrete due to repeated cycles of freezing and thawing from frost action, causing the surface to crumble. Disruption of the concrete surface can also result from salt crystallization in the concrete pores caused by repeated wetting and drying cycles during tides and storms. Finally, the time-dependent pressure on concrete subjected to wave action can be destructive.

(a)   Cracking and corrosion of concrete structures

In reinforcing concrete, tension (either direct or due to bending) causes cracks to form and/or open up. These cracks are at right angles to the force. Visible cracks should not occur in prestressed concrete. Excessive deformation due to tension can allow direct contact of the steel reinforcement with seawater through the resulting cracks, which may accelerate its corrosion. Excessive compression in concrete becomes evident by a series of progressive cracks parallel to the force. Spalling occurs as the cracks worsen. Complete failure can then follow, with only a slight increase in applied force.

Corrosion is accompanied by an increase in volume of the steel reinforcement. Initially, there may be staining on the concrete surface, following the lines of reinforcement, followed by cracking parallel to the reinforcement. Spalling then occurs, exposing the reinforcement.

(b)   Fatigue of reinforcement or prestressing tendons

This is evidenced by cracking of the concrete resulting from fatigue failure of the embedded steel, brought about by cyclic loading.

(c)   Impact damage

This is caused by the removal of the concrete cover to steel by ship collision or by heavy objects being dropped overboard.

7 Inspection and cleaning

Visual inspection is a very important and well-respected section of engineering and as such warrants a brief general introduction before concentrating on the offshore aspects of the subject.

Offshore structures represent a major engineering effort for the oil companies and as such constitute a massive investment of capital. From conception, therefore, the structures are designed and built to the highest standards so that the maximum working life can be obtained, thus maximizing the initial investment. Another equally important consideration is the very hostile environmental forces that are imposed on the structures in the North Sea. In order to withstand these forces for the entire design life of any structure it is necessary that it be built to the highest possible standards. Finally, there is the fact that a crew of men will be working and sometimes living on the structure and any failure could lead to loss of life and, in the case of a production platform also to possible massive pollution. With these factors in mind, the quality assurance and quality control exercised on offshore structures is of the highest possible standard. To gather some insight into this programme a brief outline of what is entailed will be beneficial.

7.1 OUTLINE OF QUALITY CONTROL REQUIREMENTS

The weld designers and metallurgists specify the materials to be used, the method of welding, all the welding parameters and the methods of inspection to be adopted where this is appropriate. The quality control team then has written instructions and detailed drawings to refer to, so that they can ensure the correct materials are used and the correct methods are followed. A typical inspection programme would include the following points.

7.1.1   Materials

The first job of the quality control (QC) team is to ensure that the material coming into the dock is of the required quality. For example, checking the specifications of the pipe or plate to ensure that it is to the standard required and then, once accepted, ensuring all the materials are stored in such a manner that they will not deteriorate to any appreciable extent.

7.1.2   Consumables

One of the main items here will be the welding consumables. These must be of the required type and they must be stored in the correct manner. For example, consider the case of manual metal arc welding where consumable welding rods are used. These welding rods must be stored in a clean dry environment where the temperature is controlled, for purposes of hydrogen control. This is not required for solid wires.

7.1.3   Welding inspection

Welding will be monitored by the quality control staff to ensure that the requirements of the original design specifications are adhered to. It is not possible to guarantee totally perfect welds, but this is taken into consideration by allowing tolerances throughout the welding process. This may mean that some welds go into service with some minor fabrication defects in them. If this is the case, these defects will be within specification and as such may be considered to be of no significance to the finished structure.

7.2   AREAS AT RISK

The structure goes into service having had the benefit of good design and quality control during its fabrication. During its service life, the inspection programme will be designed to monitor overall deterioration and to concentrate on areas particularly at risk.

7.2.1   Highly stressed nodes

It will be possible right from the design stage to predict those node areas on the structure which are most highly stressed. These positions will be included in the inspection programme at more frequent intervals and will be subjected to closer scrutiny than less sensitive areas.

7.2.2   Inter-tidal zone

This part of the structure is subjected to much higher corrosion rates because of the frequent exposure to wind and sea. There is an increased risk of accidental damage in this area due to the proximity of surface vessels and the fact that many items of heavy equipment are craned on board the structure.

7.2.3   Risers

These carry crude oil from the seabed reservoir up to the production platform. The crude itself may be very corrosive and may well be quite hot. Both these features will increase the corrosion rate.

7.2.4   Conductors and conductor guide frames

The drill string runs down through the conductor pipe and stresses are imposed on the conductors and on the guide frames during the drilling operations.

7.2.5   Pipelines

These are at risk in the area of the platform, particularly because of the frequent anchoring operations which go on in these areas.

7.2.6   Seabed installations

Once again, these are more at risk because of the frequency of vessels anchoring in the area.

7.3   THE SIZE OF THE TASK

Even when considering normal deterioration, the size of the task is quite formidable. Every welded joint on every nodal site on the entire structure has to be inspected periodically.

   The actual frequency of inspection is determined by how critically loaded the particular joint is. Some joints may require annual inspections, while others may be surveyed only every five years. This task will need to be divided into annual requirements. Assuming that a typical North Sea structure has some 80 nodal areas, and that some ten of these require annual inspection, 24 nodal areas will have to be inspected annually over a five year period. The weather window in the North Sea generally allows for about six months' productive diving operations. This means that four nodal areas per month, or about one per week, must be inspected to achieve this sort of inspection target. If the overall survey of the remainder of the structure is added to this programme, then the sheer size of the task begins to become apparent. Furthermore, this does not make allowances for other requirements such as routine maintenance or accidental damage.

7.3.1   Frequency and extent of inspection

Certain areas of the structure require annual inspection. As mentioned above, highly loaded nodal areas require this treatment. Riser pipes will also warrant this because of the increased risk of higher corrosion rates. Other less sensitive areas will require surveying only once every five years.

The extent of these surveys will also depend on the loading on the particular component. The highly stressed nodes will need to be inspected visually and with magnetic particle inspection (MPI) to ensure that no warning indications are missed. Less highly loaded areas will require only a visual inspection to ensure that they are sound. Sites at risk from corrosion will require cathode potential (CP) readings and may also require ultrasonic wall thickness readings. Eddy current (EMD) and alternating current potential difference (ACPD) techniques are also being used with a view to either reinforcing or superseding MPI and ultrasonic methods. In all cases, extensive use of photography and closed-circuit television (CCTV) is made to record inspection data.

7.4   PROGRAMMES FOR INSPECTION

Underwater inspection is expensive and in order to ensure that the most effective use is made of the inspection effort, detailed procedures are employed. On a structure, these procedures detail precisely what tasks are to be carried out at each level and at each location. When inspecting pipelines, detailed tasks will be contained in the procedure to ensure complete and efficient coverage. A typical procedure for an offshore structure would be as follows.

7.4.1   Typical procedure

(a) General visual inspection of the area prior to cleaning, in order to identify any obvious gross damage.
(b) Clean the inspection area down to the protective coating.
(c) Assess the condition of the protective coating and report any blisters, bare patches or areas where primer can be seen.
(d) Make a close visual inspection of welds. The weld and an area 75 mm (3 in) either side of the weld is cleaned to bare metal.
(e) Report any gross damage.

(f) Undertake a close inspection recording any defects. A tape measure fixed around the weld provides a reference.

(g) Photograph the entire inspection area.

(h) Make a video recording with the diver's commentary on any defects.

7.5   PREPARATION FOR INSPECTION

Before any structure can be inspected, items on the structure must be identified and a method must be evolved to ensure that any defects can be fixed in position so that the appropriate site can be revisited. It is also necessary to clean the structure down to bare metal to ensure an adequate inspection is achieved.

7.5.1   Marking methods

The actual system used to determine positions around a structure should be understood before the marking methods are examined.

A 'Platform North' is nominated so that orientation is possible. Then rows of legs are lettered and columns are numbered (see Fig. 7.1).

	1	2	3	4	
	0	0	0	0	A
Rows	0	0	0	0	B
	0	0	0	0	C
		Columns			

Fig. 7.1 Basic grid system.

Each leg can now be identified by referring to its column figure and row letter in a grid fashion. The position in the vertical plane can be determined by depth, which is measured from mean sea level. Figure 7.2 illustrates this.

The methods for marking structures do vary, but anti-fouling paint markings are quite common, as is the Sea Mark system. This latter method uses specially designed tiles made of copper wire and coloured yellow. The identification mark is permanently marked on the tile in black and the tile is fixed in position by bands which are tensioned up to retain the tile permanently in place.

The marking method can be illustrated by referring to Fig. 7.2. Suppose that node joint 32 on leg B4 is being photographed. The actual nodal area contains two crossbraces and two vertical diagonal braces. The actual member or weld must be identified and this can be done by reference to the brace and stating where it is connected. In this case: the crossbrace coming from 32 going to 31 on leg B3. The reference number would be written B4-32/31 and then each frame is identified by reference to the tape scale fixed around the joint.

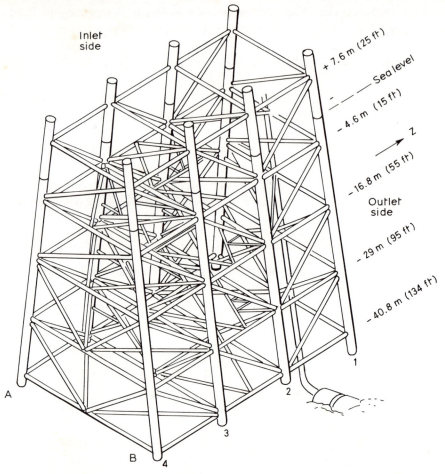

Fig. 7.2 Measurements from mean sea level.

7.6 METHODS OF CLEANING

There are basically four modes of cleaning, and within each mode there is a selection of techniques giving several methods.

7.6.1 Handtools

Tools of this type do work well, but rely very much on the enthusiasm of the diver for the best result. They are unlikely to cause any damage to the structure itself. The disadvantages are that they are slow to use; some fouling such as mussel holdfasts cannot be removed; and prolonged use leads to diver fatigue.

Examples of these tools are scrapers, wire brushes, chains and wire ropes. In use, the scrapers are the least detrimental to the surface of the structure. Chains and wire ropes are wound around the member to be cleaned and are then pulled back and forth. They can clean down to bare metal.

7.6.2   Pneumatic tools

These tools can also work very well. They are more efficient than handtools and cause less diver fatigue. The disadvantages are that below about 20 m (66 ft), they are grossly inefficient; needle guns can cause notches and thus cause damage to the structure; and grinders can be both difficult to control and remove a lot of material if not properly controlled.

Needle guns have been banned by some operators. Grit blasters have been used with some success to depths of about 30 m (98 ft). These tools can all cause damage to the surface of the structure, particularly if they are badly used. The use of these tools on concrete is limited.

7.6.3   Hydraulic tools

The advantages of hydraulic tools are the same as those of pneumatic tools, with the added advantage of no depth restriction. The disadvantages are that they are quite heavy and the hoses are stiff and awkward to handle.

Examples of these tools are grinders, rotary brushes and scrubbers. From 1985, Subsea Cleaning Systems offered a diver-controlled scrubber system which was designed to fit pipe sections specifically. The use of these tools on concrete is limited.

7.6.4   Water jets

These tools are widely used and will produce a bright metal finish in one operation. Grit entrainment is available and makes the tool even more efficient. The main disadvantages are operator safety and the fact that unskilled operators can cause damage by scouring the surface of the material to be inspected. The use of water jets on concrete is tightly controlled.

Thus, the detailed inspection programme must be determined, and the inspection sites must be thoroughly cleaned before specific areas of the structure can be inspected.

8 Visual inspection

8.1 THE SIGNIFICANCE OF VISUAL INSPECTION DURING SERVICE

When the structure comes into service, it is hoped that it is free of all significant defects. This of course depends on the rigour of the quality assurance of the fabricators. To ensure a continuous working life for the structure, it is necessary to maintain an adequate inspection programme. Such a programme must be capable of detecting possible problems at an early stage. This allows the engineers time to analyse the inspection information and suggest remedial action if required. There are, as was stated in Chapter 3, statutory requirements relating to installations in the North Sea which make it incumbent on the platform operators to obtain a certificate from government-appointed agencies stating that the platform is fit for its role. These agents require evidence from the operators that an adequate inspection programme has been carried out. The parties to whom such a programme must be acceptable have already been discussed.

Experience has shown that the vast majority of all defects found in offshore structures have been found visually. The importance of visual information is utmost both in planned visual inspection programmes and general diver observation. The training of diving personnel in visual awareness is essential and should not be left to chance if improved reliability of offshore installations is to be achieved.

8.2 THE SCOPE OF VISUAL INSPECTION

Visual inspection is concerned with the entire structure from the seabed to the top of the flare stack or derrick. At the seabed, all installations like subsea wellheads (christmas trees) or pipelines radiating from the structure, will require inspection for deterioration or damage. Interaction between installations and the seabed should be looked for, such as bridging of pipelines, scour at jacket legs, etc., and the seabed profile should be watched for subsidence and cracking. Inspection personnel should be able to carry out detailed bottom surveys to determine the profile of the seabed.

Above the mudline, any damage to the legs or braces or any of the construction welds, must be reported, identified and measured. This also applies to conductor pipes and associated guide frames, and the riser pipes of drilling platforms. The cathodic protection (CP) system, whether impressed current or sacrificial (see Chapter 9), will require close inspection to ensure that it is undamaged and that it is working efficiently. The build-up and types of marine growth will also require categorizing. Visual inspection will be the initial inspection method.

The key topics which require visual observation offshore can be summarized as: (a) marine growth; (b) welds; (c) accidental damage; (d) seabed installations; (e) movement of the seabed (scour); (f) conditions of protective coatings; (g) corrosion protection systems (anodes); (h) debris; (i) pipelines; (j) risers/conductor pipes.

8.2.1   Marine growth

Visual inspection requires information on the density of population, local density variation and the types of marine growth (in particular the presence of hard marine growth). The size of the average and largest specimen are sometimes required. Occasionally, specimens will be required for lab analysis. Usually, the method of reporting is to estimate the percentage covering of marine growth, but occasionally population counts are required. To carry out these surveys, specific areas are selected and marked out (often 30 cm² (4.7 in²)) and the precise number of each species is counted and logged. General observations such as excessive corrosion should be noted during the marine growth survey.

8.2.2   Weld inspection

After initial cleaning and preparation, the weld can be visually inspected. This will involve inspection for surface discontinuities on the weld itself and also in the heat-affected zone. Particular care will be taken to look for any signs of cracking. Corrosion damage in and around the weld will also be assessed and colours of corrosion products will be noted. The severity of pitting will be assessed and reported.

8.2.3   Damage

Concurrent with the inspection of welds, the diver inspector will be on the lookout for any type of damage to the structure. Such damage may be the result of storms, collisions, accidental dropping or deliberate jettisoning of items from the structure or supply boats, and/or stress failure.

8.2.4   Seabed installations

Usually, there will be a separate inspection programme for seabed installations, but in general, they require the same attention as a platform.

8.2.5   Scour

The amount of any scour will have to be assessed, and if severe, a profile of the seabed will be made. This can be done by using techniques like a pneumogauge or possibly with photogrammetry.

8.2.6   Condition of coatings

Where these exist, they need to be assessed for integrity. Any breaks in the coating need to be reported. If there are blisters, these need to be assessed for size and percentage covering.

8.2.7   Anodes/cathodic protection (CP) system

These are inspected and measured to assess how much anode material has been used. The devices which hold the anodes in place are checked for wear. If an impressed current system is used, its cables are inspected to ensure that the insulation is intact and that the cables are properly secured.

8.2.8   Debris

The amount, the types and sizes, the weights and the position of individual items must all be recorded. Any local damage caused by the debris should also be recorded.

8.2.9   Pipelines

The condition of the outer covering of the pipeline, any obvious damage to field joints or flanges, any missing flange bolts, any gross displacement and any free spars must all be reported. On occasion, wall thickness readings and CP readings may be required.

8.2.10   Risers/conductors

The overall condition of the risers and conductors should be noted and the condition of bracelet anodes, clamps, bolts, frames and coatings will require careful assessment. These components are particularly sensitive because of the nature of the work for which they are employed.

8.3   DIVER INSPECTION METHODS

The basic method of inspection is visual, backed up by the other senses. In this context, seeing has the following advantages:

(a) The eyes can focus from a few inches to infinity and give a wide-angle view of the area, which gives the inspector an overall perspective view.
(b) Different textures can be viewed and interpreted.
(c) Viewing is by stereoscopic vision, so depth is seen.
(d) Natural colour is seen without any artificial bias being introduced.
(e) The visual information is instantly interpreted by the brain.
(f) The actual item, and not a two-dimensional picture of the item is inspected.
(g) A verbal report can be given concurrently with the inspection.
(h) No other equipment is required apart from a light source.

The sense of touch can be used to help to determine the correct interpretation of the information before the inspector.

A discontinuity found by this method must be accurately measured and fully defined so that the fullest possible information is recorded for surface interpretation and engineering analysis. This requires the use of various other measuring techniques.

8.3.1   Methods of measuring and recording defect size

The basis of all measurement is a straight line between two points. This can be achieved in several ways, for example, by using a straightedge in the form of a ruler and reading the measurement direct or by measuring with a pair of callipers and subsequently making the measurement indirectly. Whatever the method, the defect will be defined by one of the following categories.

(a)   Linear measurements

Ruler.   Used to take small measurements such as actual crack size. Possible measuring accuracy is plus or minus 0.5 mm (0.02 in).

Tape measure.   Used for measurements up to 100 m (328 ft). Not as accurate as a ruler because of the problems of positioning the end and tape sag. Possible measuring accuracy is plus or minus 5.0 mm (0.2 in).

Magnetic tape.   Used for measurements up to 3 m (9.8 ft) and often used to take circular measurements. Possible measuring accuracy is plus or minus 1.0 mm (0.04 in).

Scales.   These are often made up with Dymo tape and are frequently used in conjunction with photography. The degree of accuracy is problematic, but with care this can be plus or minus 5.0 mm (0.2 in).

Transponders.   Sonar transponders are sometimes used for site surveys or for marking important components such as Christmas trees. Measurements of up to 1000 m (3280 ft) can be made taken with an accuracy of plus or minus 10 mm (0.39 in).

(b)   Circular measurements

Callipers are used for measurements up to 2 m (6.6 ft), but above this size the callipers are difficult to handle. Correctly used, a measuring accuracy of plus or minus 0.5 mm (0.02 in) is possible.

Special jigs.   Various jigs exist which can measure the ovality of members. Accuracy is probably of the order of plus or minus 5 mm (0.2 in).

(c)   Deformations/dents

The straightedge is used for smaller measurements; the accuracy is plus or minus 1.0 mm (0.04 in).

Profile gauge.   A variation of the straightedge method. Used correctly, a mirror image of the defect is recorded with an accuracy of better than plus or minus 0.5 mm (0.02 in).

Taut wire.   This is a common variation on the straightedge, but is used for large areas. This method can be used to measure how much out of true a member is. Measuring accuracy is problematic, but is of the order of plus or minus 5 mm (0.2 in).

Casts.   Various materials may be used to take a cast of a defect which will produce a mirror replica of that defect with exact proportions. The latest development for underwater use is produced by BP Chemicals Ltd, and is marketed as Aquaprint. The moulding agent is supplied in a cartridge which is placed into a pressure dispenser. The diver has only to pull the trigger and the system delivers a stream of material which moulds onto the surface and cures off in about 15 minutes. This flexible but non-stretchable cast can then be peeled off and removed, to be taken to the surface for analysis in the normal way.

The material can also be used to take a mould of a magnetic particle inspection. In this instance the magnetic particles are cast into the mould for a permanent, three-dimensional record.

Photogrammetry.   This is a specific application of photography which employs stereoscopic pictures of the damaged area. With current equipment, accuracy of plus or minus 2.5 mm (0.1 in) at 3 m (9.8 ft) standoff is possible. A fuller explanation of this method of measuring can be found later in this chapter.

8.4   RECORDING INSPECTION FINDINGS

Any defect which is discovered during the inspection must be recorded. Currently, methods of recording include drawings, closed-circuit television (CCTV), photography, photogrammetry and casts. All these methods are used extensively and they are all included in the final report. How they are utilized in the report is given in Chapter 15, but here the methods are discussed.

8.4.1   Drawings

These are sometimes drawn from memory after the diver has finished the inspection. More usually now, they are prepared by surface data recorders or the equivalent, at the time of inspection. When necessary, these drawings are passed to the diver for verification. However, often they are filed for collation right away. This requires an accurate record first time. This cannot be achieved without good teamwork and communications between diver and surface.

8.4.2   Closed-circuit television

As a means of recording inspection information, this method is used extensively throughout the North Sea. It is used as a handheld, diver-deployed system, or as a remotely operated, remotely placed system, mounted on remotely operated vehicles (ROVs) or used in piloted submarines. The systems available at present are monochrome (black and white), colour, and low light. Monochrome is used extensively by ROVs with colour systems being introduced regularly to replace this older method. Colour video cameras are more often used by divers in the handheld mode, with monochrome being used as head-mounted units. Low light cameras are currently used for specialist applications.

(a)   System restrictions

The problems associated with television all revolve around the basic principles of how the system works, and are caused by the design limitations of the system. As stated above, the television camera only transmits a series of still pictures, and there is a fixed speed of one field every 1/50 second. This means that the system is incapable of producing a well-resolved picture if the relative movement between camera and subject is above a certain speed. Thus, if the camera is traversed over the surface of an object quickly, the surface will appear to be blurred when viewed on the monitor. If any fast movement at all occurs in front of the camera, the picture will appear blurred. The diver, for example, can be affected by the tide, and if he is moved quickly by the current, then so is the camera. Slow camera movement is therefore of paramount importance.

The picture itself is made up of a series of lines (625 lines per frame), and this has the effect of limiting the amount of fine detail that can be recorded. This limit on picture quality may not be noticeable on the monitor, particularly with a colour system. Nevertheless, the system is incapable of resolving fine detail. This, coupled with the fact that the system shows two-dimensional pictures, means that the picture viewed on the monitor should not be accepted entirely at face value.

Finally, it must be remembered that the camera target is automatically changing light energy into electrical energy. This can lead to the problem of flare, and it is possible that strong light entering the camera can burn out the target altogether. Flare occurs when the camera lens is pointed at a scene which consists of a large, quite bright area and a contrasting area which is dark or in shadow. The target is incapable of resolving the two opposites and therefore all detail disappears in the dark area, which then appears black. The same occurs in the bright area, which then appears to glow very brightly indeed.

To record inspection information, the camera signal is recorded on a video tape recorder (VTR) or a video casette recorder (VCR). This entails using the electronic signal from the camera to produce a magnetic field which affects a magnetic tape in the tape recorder. Electrical energy is changed into magnetic energy, and there will be losses of energy which will result in a loss of picture quality.

Currently, these restrictions have in one way or another led to a preference for photography as the prime method of recording inspection information. However, developments continue, and these may change this situation in the future.

(b)   Video recording

In the oil industry there are two methods of recording video inspections: video home system (VHS) and U-Matic. The VHS equipment is smaller, and more portable and the equipment is more universally available. There is good compatibility between different manufacturers' equipments. The main restrictions are that reproduced picture quality can be poor and if copies of the tapes are required these copies are always of noticeably poorer quality. The U-Matic system uses a much wider tape which therefore records in much greater detail. Picture quality is therefore much better and if copies of tapes are required then reproduction is also much better. In either case, recordings are made selectively to avoid the necessity of spending many hours reviewing tapes. One important aspect of video recordings is the ability to include a sound commentary, which may be either the diver's comments or views recorded by surface personnel.

These commentaries themselves can be quite important, especially when they come from the diver. After all, it is only the diver who has seen the actual item and can view it in stereoscopic vision. To avoid the possibility of any errors, the diver should follow a set procedure when making a commentary. This need not be too rigid, but the diver should give:

  (i) name
 (ii) date
(iii) location
(iv) description of object

When the diver is breathing a helium mixture, the voice changes pitch and it becomes necessary to use special voice unscramblers. In these circumstances, the diver must speak very carefully indeed to ensure that it is possible to understand what is being said.

The sound commentary produced on tape then should be of considerable interest when the recording is reviewed, because the diver will be describing what is being viewed. In some cases, it may be that the diver's comments are overrecorded by a prepared commentary. Even when this occurs, the original soundtrack will have provided much useful information.

(c) Image enhancement

Image enhancement is becoming a fairly normal technique with video broadcasting, and the necessary equipment is becoming more widely available. At present, the technique is not being used offshore, but if the need for it arises, it will surely be introduced quickly. One point to note is that extremely accurate and very finely detailed still photographs can be taken with very little effort and this may eclipse any need for image enhancement.

8.4.3   Photography

Extensive use is made of photography to make a permanent record of the inspection. Colour film is normally used in order to make interpretation more straightforward.

Before examining photography in more detail, however, it will be useful to outline the effects that water has on light. The film in the camera will not react properly, and therefore will not record a proper image, unless the correct amount of light is exposed onto it. Light is all important in this context and it is essential to have fixed clearly in one's mind the factors affecting light in water.

(a)   Reflection and refraction

As light hits the surface of the water, it is subjected to reflection and refraction at the air/water interface. Reflection reduces the total amount of light going into the water and refraction causes dispersion of the light.

(b)   Loss of intensity

Light in water is absorbed quite quickly because of the dense nature of the water. This absorption is normally described as attenuation. Thus, some of the light energy is transferred to the atoms of the water as the light rays pass. The light intensity is further reduced by the effects of scatter, where light striking suspended particles in the water is reflected in various directions off the path of travel (see Fig. 8.1).

(c)   Loss of colour

Some comments on light and colour are contained in Section 8.4.2. The various colours of the spectrum have different wavelengths, and as white light penetrates the water, different colours are filtered out because various wavelengths are absorbed by the water. The longer wavelengths are the first to be filtered out because they are less intense and therefore have less penetrating power than the shorter wavelengths. Figure 8.2 summarizes these effects on light penetrating pure seawater.

(d)   Loss of contrast

When we look at an object, its form and texture are made apparent to our eyes by the different intensities of reflected light coming off the surface of the object. If the difference in the intensities is marked then the contrast is good and the image is said to be 'crisp'. If the differences are poor then contrast is poor and the image is said to be 'muddy'. In water, there is always less light available than in air, and as depth of water increases, so available light decreases. This means that in water there is always poor contrast and this situation becomes worse with depth.

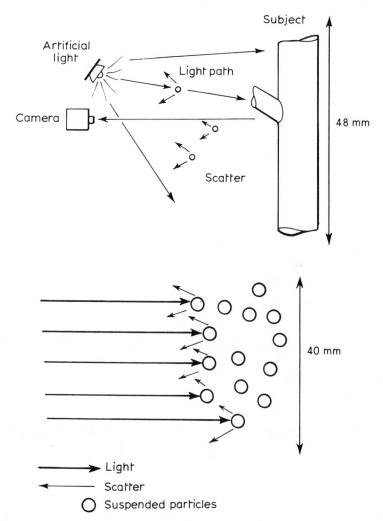

Fig. 8.1   Scatter.

The result of all these factors affecting light is that photography using available light under water is only really practicable at shallow depths. Deeper than this, it is only possible to get good quality results on a sunny day when the sun is nearly vertical, say between 10 am and 2 pm, and when the water is clear, with no suspended particles. Finally, it should preferably be either in the tropics or at least during summer, when the sun is more vertically overhead (see Fig. 8.1). These restrictions on the use of photography using available light usually mean that artificial light in the form of a flash unit is used instead.

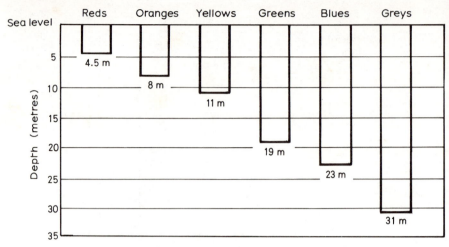

Fig. 8.2   Absorption of colours from the light spectrum with increasing depth.

8.4.4   Flash photography

Currently, flash photography is generally used extensively in oilfield diving. There are three major variations on the way that the flash is deployed. Firstly, as a single unit on a stalk or arm. Secondly, two units are deployed on either side of the camera, each on its own arm. Thirdly, the flash is built around the camera lens as a ring flash. This last application is the least flexible and tends to be used with specialized close-up cameras. In the case of single or double units, the two major problems to be dealt with are back scatter and reflected light from the object.

(a)   Back scatter

This effect is caused by light being reflected from suspended particles in the water. Its effects can be minimized by placing the flash to one side (see Fig. 8.3).

(b)   Reflected light

The problem of reflection can be minimized by carefully placing the flash at an angle, as indicated by Fig. 8.3. When flash photography is used, the camera controls are set in accordance with the manufacturer's instructions.

8.4.5   Film stock

The film inside the camera records the image which has been illuminated by either the available light or by the flash. A few comments on the type of film being used offshore will be of interest.

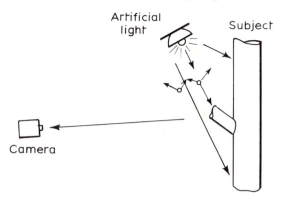

Fig. 8.3 Avoiding back scatter.

Before looking at this film reaction in more depth, let us consider some other aspects of selecting film type, bearing in mind that all films react to light in the way just described. The type of film in use can be selected by determining whether colour or monochrome is required and also if small, medium or large format is needed. Similarly, special films such as infrared can be selected.

The type of film for colour or monochrome can be selected from the following:

(a) Monochrome films are traditional films using silver halides to record the image, which is then printed. Modern monochrome film converts the silver halides to dyes in the processing. This gives a finished print which has no discernible grains.

(b) Colour films should be split down further into colour prints and colour slides.

(c) Colour prints (colour negatives) are films which convert the silver halides into dyes to record the image as colour opposites.

(d) Colour positives (slides) are films which convert the silver halides into dyes which record the image in true colours ready for viewing without further processing.

In general, for underwater photography, there is a choice of two film formats:

(i) Small format: where the negative size is 35 mm × 24 mm and the film stock is referred to as 35 mm.

(ii) Medium format: where the negative size can be 6 cm × 6 cm (60 mm × 60 mm) or 6 cm × 7 cm (60 mm × 70 mm). The most popular film stock for this is 120 roll film.

As can be seen, the selection of a particular film type depends on three factors: how quickly it reacts to light, whether it is colour or monochrome, and whether it is small or medium format. The first consideration must be its reaction to light, which is referred to as the 'film speed'.

In order to have an integrated system for photography which will quantify reaction to light in a way which is logical to use with a camera, special numerical systems have been evolved. These systems give numerical values to the speed of reaction to light of the film and these numerical values have certain relationships with each other. The first is that as the film reacts faster the value is higher; for example, 50 ASA is faster than 25 ASA. Secondly, as the value increases by one step, the light reaction increases precisely twofold; for example, 100 ASA is twice as fast as 50 ASA. This equates to one full stop on the camera aperture or speed control.

Films are loosely grouped into slow, medium and fast films, according to the ASA rating. These points are summarized in Fig. 8.4, which selects the main film speeds in the ASA (American Standards Association), DIN (Deutsch Industries Norm) and ISO (International Standards Organization) systems: 25 and 50 ASA are slow films; 100 and 200 ASA are medium; 400 and 800 ASA are fast. Each film is twice as fast as the previous one in value. For example, 100 ASA is twice as fast as 50 ASA but only half the speed of 200 ASA.

System	Film speeds					
ASA	25	50	100	200	400	800
DIN	15	18	21	24	27	30
ISO	25/15	50/18	100/21	200/24	400/27	800/30

Fig. 8.4   Rating systems and film speeds.

It is worth looking in a little more depth at this reaction to light, because of its vital importance to photography. The reasons for films reacting to light more or less quickly is because of the size of the grains of silver halides on the emulsion. The smaller the grain the slower it reacts to light. By selecting grains of all the same size, the film manufacturers determine its ASA rating. The other important quality which is also determined by grain size is the finished print quality. The larger the grain the poorer the quality. Thus, when selecting a film type, it becomes a trade-off between light reaction and print quality.

The selection of film type with regard to colour or monochrome really comes down to a choice of selecting between a lifelike image or not, and virtually no monochrome film is used offshore. This means a decision between colour prints or colour positives. The advantages and disadvantages of these films can be summarized as follows.

Colour positives' advantages are that they can be viewed without having to process prints and that they are easy to develop on site and quickly available (typically, 45 minutes). Enlargements are also easy to obtain using slide projectors. However, prints are difficult to make from slides, and slides cannot be presented in a report. Also, this system of photography is intolerant of exposure errors.

Colour prints are tolerant of exposure errors and enlargements are easy to make to any size. Reprints are easy to obtain and they are easily presented in reports. It is also easy to study details on prints. However, their processing requires more time and equipment.

Finally, there is the choice between small and medium format. This is a straightforward choice of print quality and is summarized in Fig. 8.5.

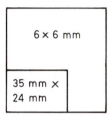

Fig. 8.5   Small and medium format.

For any size print, the degree of enlargement for the 6 cm × 6 cm (2.4 in × 2.4 in) is much less than for the 35 mm. This means less loss of quality in the final print and therefore much greater detail. Figure 8.6 summarizes this by selecting a particular film stock for a particular application. In all cases, whenever possible use colour negative film.

8.4.6   The camera

The camera is basically a lightproof box with a hole called the aperture, which allows light in from one point only. In its simplest form, a shoe box with a pinhole in one end would serve as a camera.

The types of camera in general use in underwater photography are, however, more sophisticated than this. Lenses are used to focus the image inside the camera and the light input is controlled with shutters and diaphragms which are

Requirement	Type of film	Suggested ASA rating
Flash photography	Slow/medium	50 or 64
Available light photography	Medium/fast	200 or 400
Requirement for good enlargements	Medium format 6 cm × 6 cm	Depends on type of light

Fig. 8.6   Selection of film.

both variable. An understanding of how the shutter and diaphragm are interrelated will certainly prove helpful and the following information should provide that understanding.

The amount of light entering the camera is obviously determined by the size of the aperture and this is controlled by the diaphragm. The quantity of light entering the camera is also determined by the length of time the aperture remains open and this is controlled by the shutter. On the type of equipment being considered here, the diaphragm and shutter can be set to specific values by controls on the camera or the lens itself. By careful adjustment, the amount of light entering the camera can be precisely controlled to suit the available light conditions and the film in use.

The size of the aperture is set by adjustment of the aperture control which is calibrated in 'F' stops. The 'F' stands for Factor and it is determined by the following equation:

$$F \text{ stop} = \frac{\text{Focal length}}{\text{Aperture diameter}}$$

Because the F stops are all factors, the relationship between them is not obvious. The difference in light units between one F stop and the next is a ratio of 1:2. This is explained by Fig. 8.7, which shows a typical F stop series for a typical camera.

F stops	2.8	4	5.6	8	11	16	22
Light units	32	16	8	4	2	1	1/2

Fig. 8.7   F stops.

F4 allows twice the light of F5.6 to enter, but only half the light of F2.8. F5.6 allows twice the light of F8 but only half the light of F4, etc.

The shutter speed is set by adjusting the shutter speed control, which is calibrated in seconds and parts of seconds. Here, the relationship between stops is quite obvious and is the same as for F stops and is the same as for film speeds. A typical series is shown in Fig. 8.8.

1   1/2   1/4   1/8   1/15   1/30   1/60   1/125   1/250   1/500   1/1000

Fig. 8.8   Shutter speeds.

1/2 second allows only half the light of 1 second to enter but twice the light of 1/4 second, etc.

The relationship between F stops and shutter speeds should now be obvious.

(a) They both control light entering the camera.
(b) They are both calibrated.

(c) The calibration is such that the light entering the camera is affected in the same way and by the same amount by both controls even though they are calibrated differently.

Before going further, there is an important feature concerned with the aperture which has not yet been considered. This is concerned with depth of field.

If one considers a subject at a distance (*d*) away and one focuses the camera on that subject correctly, the image of it will be recorded correctly in focus. But consider the area in view in front of the subject (*a*) and behind the subject (*b*). See Fig. 8.9.

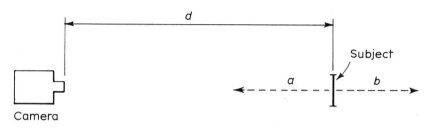

Fig. 8.9   Depth of field.

Both these areas can be recorded either in or out of focus depending on the F stop in use. These areas together make up the depth of field. The criterion is the smaller the aperture, which means the higher the F stop, the greater the depth of field and vice versa. Figure 8.10 should be of help in understanding this feature of the camera. With the camera focused to 1.5 m (4.9 ft), the depth of field increases as the aperture is stepped down; i.e., at F2.8 only, the subject and 0.1 m (0.33 ft) of the foreground and background are in focus, but at F22, the

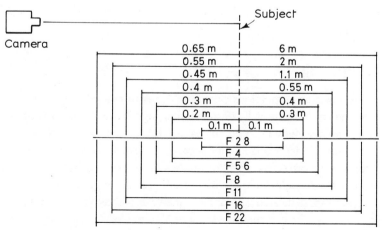

Fig. 8.10   Aperture and depth of field.

subject and 0.65 m (2.13 ft) of foreground and 6 m (19.69 ft) of background are all in focus.

Given any particular film speed, a certain quantity of light will be required before an image is recorded. The slower the film the greater the quantity of light required. As has been shown, the quantity of light is monitored by the aperture and the shutter. If the two controls are considered together, it should be apparent how the two interreact in practice.

If the exposure value required is such that F8 at 1/60 second will pass enough light, F5.6 at 1/125 second and F11 at 1/30 second will also pass the same amount of light (see Fig. 8.11).

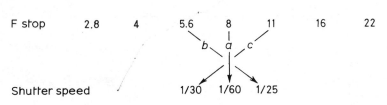

Fig. 8.11   Aperture and shutter speed.

This gives the background to the camera controls and, for more detail, individual instruction books for each camera will need to be consulted.

8.5   TYPES OF CAMERA

Currently in the North Sea, there are several specialized companies offering photographic services to the offshore industry. This specialist assistance has created a mixture of camera types with different systems being available for different tasks.

8.5.1   The Nikonos system

The Nikonos is an amphibious camera which is totally waterproof by design, with no need to have specially designed housings. It is a unique type of camera, being a simple viewfinder type but having interchangeable lenses. There is a variety of accessories available for the camera, making it a very flexible system. It is possible to take close-up photographs using either adaptors or close-up lenses and it is equally possible to use either wide-angle or even telephoto lenses. There are dedicated flash units available as well as a wide variety of non-dedicated units. The camera uses standard 35 mm film cassettes and because of its simplicity it is widely used.

8.5.2   Housed 35 mm cameras

To achieve better quality pictures it is possible to take a good quality 35 mm land camera and design a housing to go around it. This has been successfully done with a good many types of single lens reflex camera. There is another advantage with this system in that through-the-lens viewing is possible so that there is little chance of not getting the correct subject in the frame. With some cameras of this type, it is possible to use bulk-loaded film, and up to 300 frames can be loaded at one time.

8.5.3   Purpose-made cameras

Purpose-designed 35 mm underwater cameras are also used frequently, and the Hydroscan is currently the most widely used of this type. This system has been designed specifically to produce close-up photographs, and it provides a close-up lens, fixed prods to achieve the correct standoff, a ring flash, 250 frame bulk film cassette and an automatic film advance and purpose-built underwater housing. Variations on this design also provide digitalized electronics which can print simple film information on each frame.

8.5.4   Medium format cameras

The majority of these systems employ cameras based around the Hasselblad format of 6 cm × 6 cm (2.4 in × 2.4 in), single lens reflex equipment. In fact, several systems exist which are exactly Hasselblad cameras in housings. This equipment has the advantage of providing a very large negative which therefore gives a very good quality print. It also gives all the advantages of through-the-lens viewing and framing. Bulk packs are available to provide a greater number of exposures.

8.6   PHOTOGRAMMETRY

This specialized application of photography has been available for use under water for some time now. Photogrammetry is the analysis of the information contained in two pictures of the same scene taken from different angles. The basis of the system is stereo photography. It is possible to take stereo pictures of a scene with a camera specifically designed to have two lenses. Each shutter is synchronized so that each time the shutter release is pressed, the camera records two frames, one in each lens (see Fig. 8.12).

This type of camera is only really effective on land, and for underwater use, the method of producing the stereo pictures is to employ two cameras. To maximize the amount of information recorded on the film and to ensure the

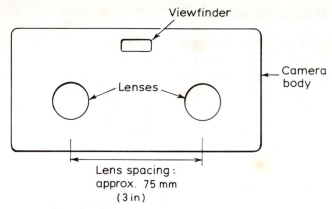

Fig. 8.12   Front view of a stereo camera.

highest degree of accuracy, these cameras are usually based on the Hasselblad 6 cm × 6 cm (2.4 in × 2.4 in) equipment.

The individual cameras are calibrated before use to match each one to the computer model of an ideal camera by photographing special calibration frames. This procedure ensures that each camera is matched to its partner when they are paired up in use. The system normally employs calibration devices in use as well, to allow for different camera angles and water conditions. These devices may appear in each frame or may be photographed only periodically through the film. Once the stereo pictures have been recorded, they can be analysed as necessary by computer. The computer printout can be in a variety of forms to suit individual requirements, and will interface with other computer systems by arrangement.

The main asset of the system is the fact that very accurate measurements can be obtained direct from the photographic record. The degree of accuracy may be of the order of plus or minus 0.3 mm (0.01 in), and measurements in all three dimensions can be obtained from the one pair of photographs. This branch of photography is likely to expand much more in the future because of the ability to obtain these very accurate results.

8.7   VISUAL DEFECTS IN WELDS

An inspector must be capable of recognizing a fault in a weld and subsequently be able to describe this fault accurately. With this in mind, the Welding Institute in Great Britain uses a standard set of phrases which exactly describes all sorts of faults in welds. These terms are used throughout the welding and inspection industries and are also recognized internationally. These standard phrases themselves are based on those given in BS 499: 1965 Welding Terms and Symbols – Part 1, Welding, Brazing and Thermal Cutting Glossary and Part 3,

Terminology of and Abbreviations for Fusion Weld Imperfections as Revealed by Radiography.

The following is a series of diagrammatic descriptions of faults which give surface indications of their presence (Figs 8.13–8.30). Stress caused by loading or accidental damage is the cause of the majority of the flaws that the diver-inspector will see, as defects produced during construction should have been identified and the serious defects corrected during the routine inspection carried out at the manufacturing stage. The aim of giving this abridged list of weld defects is to introduce readers to some of the more common faults and to start to build up their welding vocabulary.

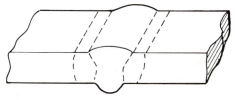

Fig. 8.13   Excessive penetration.

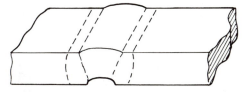

Fig. 8.14   Incomplete penetration.

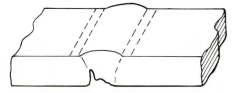

Fig. 8.15   Lack of root fusion.

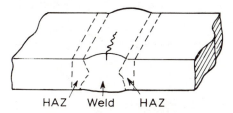

HAZ   Weld   HAZ

Fig. 8.16   Centre line cracking.
            *heat affected zone

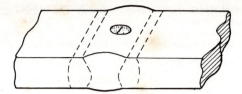

Fig. 8.17   Crater crack.

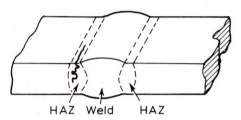

HAZ   Weld   HAZ

Fig. 8.18   Lamellar tearing.

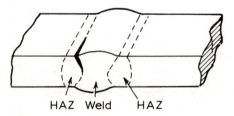

HAZ   Weld   HAZ

Fig. 8.19   Lack of sidewall fusion – where the surface indication is a toe crack.

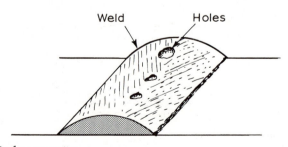

Weld   Holes

Fig. 8.20   Surface porosity.

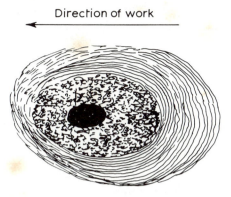

Fig. 8.21 Crater pipes.

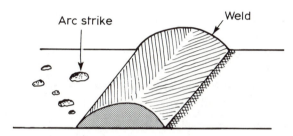

Fig. 8.22 Arc strikes – where the electrode is accidentally struck at random on the surface of the metal.

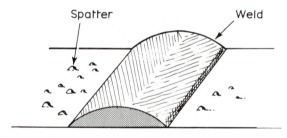

Fig. 8.23 Spatter.

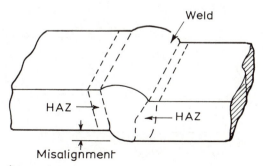

Fig. 8.24 Misalignment.

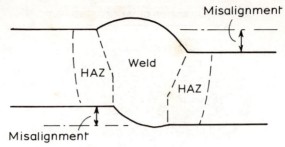

Fig. 8.25 Cross-section of misalignment.

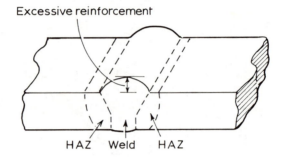

Fig. 8.26 Excessive reinforcements.

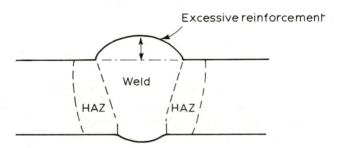

Fig. 8.27 Cross-section of excessive reinforcement.

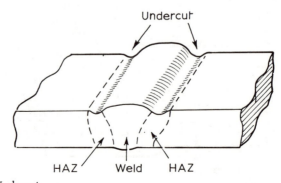

Fig. 8.28 Undercut.

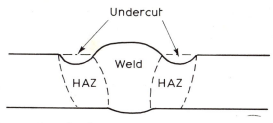

Fig. 8.29   Cross-section of undercut.

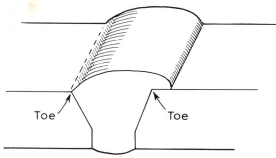

Fig. 8.30   Overlap.

For more detailed information, BS 499: 1965 Parts 1 and 3 should be consulted. The Welding Institute and various other interested parties also produce some excellent literature on this subject. However, the best way to learn the appropriate welding terms is to use them in practice.

8.8   REMOTELY OPERATED VEHICLE (ROV) INSPECTION

The role of the ROV is expanding as technology advances. The advantages of this form of inspection are:

(a) The ROV is not limited by decompression restrictions. It can be operated at any depth from seabed to surface and can be moved at will from one level to another at the touch of a button.
(b) It can provide a very stable camera platform.
(c) Horizontal movement is largely unrestricted and is achieved without the jerky movements and stops associated with a diver doing a similar manoeuvre.

The ROV is used to inspect pipelines, concrete structures, steel structures for damage, and seabed installations, and to take cathode potential (CP) readings, wall thickness readings, stereoscopic pictures for photogrammetry, and still pictures.

In all these roles, the prime method of inspection is video. Monochrome and colour video cameras are mounted on the vehicles and the pictures are monitored on the surface. This does mean that the video image and not the object itself is inspected, which in turn means that fine detail cannot be observed. This is of little consequence when the item under inspection is a large offshore concrete production platform. In this case, hectares of concrete have to be inspected and the ROV is a most useful tool. If a close weld inspection is required, however, then fine details can only be seen if photographs can be taken. The ROV does have access limitations and this may mean that the task is beyond the scope of the vehicle. The taking of wall thickness readings relies on the skill of the operator in 'flying' the ROV close enough to touch the structure with a probe. Again, access problems can prevent readings being taken. CP readings may also require contact, but they may be obtained by proximity readings (see Chapter 9).

In the future, systems may be produced to undertake even the detailed inspection of welds and the results of such inspections may be interfaced directly onto computers. At the present, however, such development is some way off.

8.9   INFORMATION OBTAINABLE FROM VISUAL INSPECTION

The actual information which can be obtained from visual inspection comes under two major categories. The first is quality assurance. Information gathered from an inspection which confirms there are no defects and that the structure appears to be intact will reassure the designers and structural engineers as to the rate of deterioration and that the overall stresses are performing within design limits. Negative findings in this case can be used to help confirm that the design criterion was correct and has not been exceeded. An ongoing inspection programme also gives confidence because it can safely be assumed that any serious defects should be detected before they become catastrophic.

The other category is defect management. With sound and accurate information available to the engineers, it becomes a practical proposition for them to monitor defects when they are of a non-critical nature. This could mean significant savings in time and resources and therefore in money, without any significant loss of safety.

8.10   AN OFFSHORE INSPECTION, AUTUMN 1986

8.10.1   Inspection requirements

The client, a major offshore contractor, required a visual inspection of an offshore oilfield. The specific requirements were:

(a) Visual inspection of one of the seabed wellheads, assessing:
   (i) Damage and debris.
   (ii) Corrosion.
   (iii) Marine growth.
   (iv) Condition of coatings.
   (v) Condition of anodes.
   (vi) Selected cathode potential (CP) readings.
(b) Video inspection of the flowlines connecting the wellhead to the single articulated leg support (SALS) base over a predetermined distance, assessing:
   (i) Damage and debris.
   (ii) Condition of coatings.
   (iii) Free spans.
   (iv) It was also required that a plot of the actual position of the flowline on the seabed be confirmed.
(c) Video inspection of the SALS base itself, assessing:
   (i) Damage and debris.
   (ii) Corrosion.
   (iii) Marine growth.
   (iv) Condition of anodes.
   (v) Selected cathode potential (CP) readings.
   (vi) Actual position and condition of all flowline and control line connections.
(d) Video inspection of the universal joint connection between the SALS base and its riser, assessing:
   (i) Damage and debris.
   (ii) Marine growth.
   (iii) Corrosion.
   (iv) Selected cathode potential (CP) readings.
   (v) Condition and electrical integrity of the earth continuity straps.
(e) Video inspection of the riser, assessing:
   (i) Damage and debris.
   (ii) Marine growth.
   (iii) Corrosion.
   (iv) Condition of anodes.
   (v) Selected cathode potential (CP) readings.
   (vi) Condition of coatings.
(f) Video inspection of the hull of the fixed position storage unit (FPSU), assessing:
   (i) Damage and debris.
   (ii) Marine growth.
   (iii) Corrosion.
   (iv) Condition of anodes.
   (v) Selected cathode potential (CP) readings.

    (vi) Condition of coatings.
    (vii) Condition of and amount of fouling on the seawater inlets and outlets.

8.10.2   Inspection method

The entire survey was undertaken with an ROV which had SIT (silicone intensified target) and colour video cameras as well as a still camera mounted on it. The survey report consisted of a written report compiled from the video, the video recordings themselves, and the still pictures which had been taken of areas of interest.

9 Corrosion

9.1 ENERGY CONSIDERATIONS IN CORROSION

With time, most materials react with their environment to change their structure. In metals it is called corrosion; in polymers (plastics), degradation; in concrete it is referred to as weathering.

In metals, corrosion is defined as the chemical or electrochemical reaction between a metal and its environment which leads to one of three consequences:

(a) The removal of the metal.
(b) The formation of an oxide.
(c) The formation of another chemical compound.

This change in the metal will be expected if we consider the energy state of the materials used (see Fig. 9.1). Generally, the metals we use, such as iron and aluminium, exist in the natural state as chemical compounds with oxygen, carbon, etc., and we extract the metal from the ore by the addition of energy in a smelting or electrolytic process. The final metal produced is therefore at a higher energy level than the ore from which it was made.

One of the fundamental laws of equilibrium is that all systems try to reduce their energy level to a minimum. This is why water runs downhill. In the same way, metals tend to reduce their energy by forming lower energy compounds. This is called corrosion, or in the case of steel, rusting. The metal reduces its energy by forming a chemical compound like the oxide.

9.2 THE CORROSION PROCESS

Having seen that there is a driving energy for the process, we now need to consider a mechanism by which corrosion can take place. In order to do this, it is necessary to remind oneself of the basic structure of the atom. In its simplest

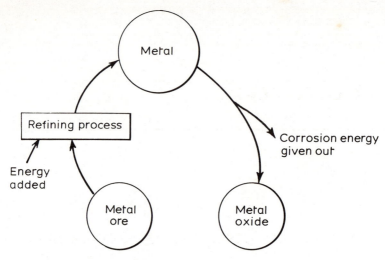

Fig. 9.1 Changes in energy levels.

form (see Fig. 9.2), it is a positive nucleus surrounded by negatively charged electrons.

The overall charge of the atom is zero and an atom is so composed that the negative charge of the electrons is equal to the positive charge of the nucleus. However, electrons can be added to or taken away from the group which surrounds each atom. When this happens, the overall charge on the atom is no longer zero. This condition of the atom is called 'ionic'. Thus, if the atom loses an electron, it becomes a positive ion, which means that the atom now has a positive charge. If it gains an electron, it becomes a negative ion; that is, the atom now has a negative charge.

The first step in the corrosion process is that metal atoms change their state

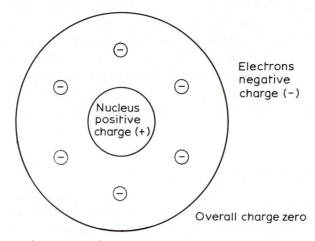

Fig. 9.2 Simple structure of an atom.

from being metallic (no charge on the atom) to being ionic (having a charge on the atom), by losing an electron from the outer shell. Corrosion is sometimes defined as the translation of metal atoms from the metallic to the ionic state. The process of corrosion goes on at the atomic level, each atom losing one or more electrons to become an ion.

The reaction in which the metal is changed from its metallic state into its ionic state is known as an anodic reaction and is shown in Fig. 9.3.

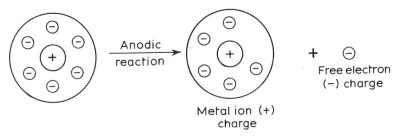

Metal ion (+)
charge

Fig. 9.3   Anodic reaction.

The site at which it takes place is the anode. The reaction at the anode for iron is that the iron atom loses two electrons (see Fig. 9.4).

This is one part of the reaction in electrochemical corrosion which takes place in the presence of an electrolyte, which is often water or a water-based solution of ionic compounds such as acids, bases or salts. The metal ion passes into solution and the electron passes through the metal which is not actually being corroded; that is, an electric current flows (see Fig. 9.4). These 'free' electrons formed in the anode reaction must be 'used up' if the reaction is to proceed. The other part of the reaction in the electrochemical corrosion process therefore takes place at the site where the free electrons are neutralized and is known as the cathodic reaction. This reaction takes place as indicated in Fig. 9.5.

A typical reaction is for the free electrons to be taken up by positive ions in the electrolyte and atoms of oxygen (see Fig. 9.6). This gives the oxygen a

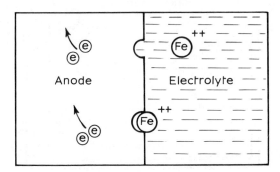

Fig. 9.4   At the anode, the iron atom loses two electrons and passes into solution.

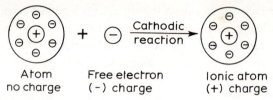

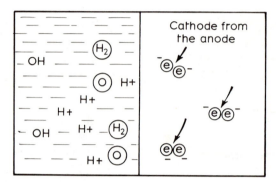

Fig. 9.5   Cathodic reaction.

Fig. 9.6   Free electrons move through the metal to the surface of the cathode to combine with positive ions and cause negative ions in the electrolyte.

negative charge. Oxygen readily accepts the free electrons because for its electron stability it needs eight electrons in its outer valency shell.

We now have positive metal ions formed at the anode and negative oxygen ions formed at the cathode so that the conditions are favourable for these to be attracted to each other and form an oxide. This ionic attraction is the same type of attraction as that which is found between sodium and chlorine to form sodium chloride (common salt); sodium is the positive ion and chlorine the negative ion.

The corrosion process therefore requires a circuit (see Fig. 9.7) consisting of an anode, at whose surface the metal is ionized, connected (electrically) to a

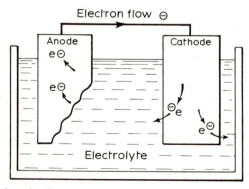

Fig. 9.7   Corrosion circuit.

cathode, where the electrical balance is restored as the electrons pass into the electrolyte. All electrochemical corrosion takes place by setting up cells like this. The size of cell will vary from that set up under one droplet of water on a metal sheet, to a complete metal platform.

9.3   CORROSION CELLS

Corrosion cells, using the corrosion process just described, can be set up by many different means, but they all operate because there is some dissimilarity between the anode and the cathode, such as dissimilar metals, dissimilar phases in the grains of the metal, dissimilar energy levels between the grain and the grain boundary of the metal, or dissimilar ion concentrations or oxygen concentrations. We will now consider some of these.

9.3.1   Dissimilar metal corrosion cell

It is found that when dissimilar metals are placed in the same fluid (electrolyte) a potential difference (voltage) exists between them. This can be demonstrated easily by placing two rods of different metals in water and connecting a voltmeter between them. The voltmeter measures a voltage (see Fig. 9.8) and if connected, a current (electrons) flows from the anode to the cathode via the outside connection. The cell acts as a very basic, low-powered battery and in battery terms the anode is the negative and the cathode the positive. Electrons flow from the negative terminal to the positive terminal in the external circuit.

Under standard conditions, where the electrolyte is dilute sulphuric acid at a temperature of $25°C$, the potential of various metals is measured and given in a

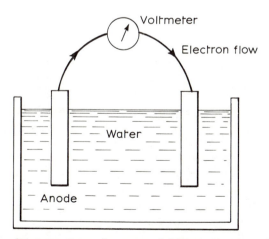

Fig. 9.8   The voltmeter registers the potential difference between the anode and the cathode.

table known as the Electrochemical Force Series or Electromotive Series (see Fig. 9.9). From the table, it will be seen that any metal will be anodic to any metal lower in the table and cathodic to any metal higher in the series. However, this table applies only to the standard conditions.

Similar tables are produced for metals under actual conditions and these are called Galvanic Series. Figure 9.10 gives the arrangement of metals in seawater.

The same rule applies to the Galvanic Series; i.e. metals found lower in the series are cathodic to any metal above them, and metals found higher in the series are anodic to any metal below them. For example, zinc is higher in the series than mild steel; therefore, if zinc is connected to mild steel and immersed

Metal atom	Electrode reaction atom to ion	Potential in volts standard electrode at 25°C
Potassium	$K \rightarrow K^+ + e^-$	−2.92
Calcium	$Ca \rightarrow Ca^{++} + 2e^-$	−2.87
Sodium	$Na \rightarrow Na^+ + e^-$	−2.71
Magnesium	$Mg \rightarrow Mg^{++} + 2e^-$	−2.34
Beryllium	$Be \rightarrow Be^{++} + 2e^-$	−1.70
Aluminium	$Al \rightarrow Al^{+++} + 3e^-$	−1.67
Manganese	$Mn \rightarrow Mn^{++} + 2e^-$	−1.05
Zinc	$Zn \rightarrow Zn^{++} + 2e^-$	−0.76
Chromium	$Cr \rightarrow Cr^{+++} + 3e^-$	−0.71
Gallium	$Ga \rightarrow Ga^{+++} + 3e^-$	−0.52
Iron	$Fe \rightarrow Fe^{++} + 2e^-$	−0.44
Cadmium	$Cd \rightarrow Cd^{++} + 2e^-$	−0.40
Indium	$In \rightarrow In^{+++} + 3e^-$	−0.34
Thallium	$Tl \rightarrow Tl^+ + e^-$	−0.34
Cobalt	$Co \rightarrow Co^{++} + 2e^-$	−0.28
Nickel	$Ni \rightarrow Ni^{++} + 2e^-$	−0.25
Tin	$Sn \rightarrow Sn^{++} + 2e^-$	−0.14
Lead	$Pb \rightarrow Pb^{++} + 2e^-$	−0.13
Hydrogen	$H_2 \rightarrow 2H^+ + 2e^-$	0.00
Copper	$Cu \rightarrow Cu^{++} + 2e^-$	0.34
Copper	$Cu \rightarrow Cu^+ + e^-$	0.52
Mercury	$2Hg \rightarrow Hg_2^{++} + 2e^-$	0.80
Silver	$Ag \rightarrow Ag^+ + e^-$	0.80
Palladium	$Pd \rightarrow Pd^{++} + 2e^-$	0.83
Mercury	$Hg \rightarrow Hg^{++} + 2e^-$	0.85
Platinum	$Pt \rightarrow Pt^{++} + 2e^-$	ca.1.2
Gold	$Au \rightarrow Au^{+++} + 3e^-$	ca. 1.42
Gold	$Au \rightarrow Au^+ + e^-$	1.68

Fig. 9.9   Electrochemical Force Series.

Magnesium	Tin
Magnesium alloys	Muntz metal
Zinc	Manganese bronze
Galvanized iron	Naval brass
Aluminium 52Sh	Nickel (active)
Aluminium 4S	78% Ni, 13.5% Cr, 6% Fe (active)
Aluminium 3S	Yellow brass
Aluminium 2S	Admiralty brass
Aluminium 53S-T	Aluminium bronze
Alcad	Red brass
Cadmium	Copper
Aluminium A17S-T	Silicon bronze
Aluminium 17S-T	5% Zn, Ni, Bal, Cu
Aluminium 24S-T	70% Cu, 30% Ni (Monel)
Mild steel	88% Cu, 2% Zn, 10% Sn
Wrought iron	88% Cu, 3% Zn, 6.5% Sn, 1.5% Pb
Cast iron	Nickel (passive)
Ni-resist	78% Ni, 13.5% Cr, 6% Fe (passive)
13% chromium stainless steel (active)	70% Ni, 30% Cu (Monel)
50-50 lead tin solder	18-8 stainless steel (passive)
18-8 stainless steel (active)	18-8 3% Mo stainless steel (passive)
18-8, 3% Mo stainless steel (active)	Silver
Lead	Gold

Fig. 9.10   Galvanic Series showing the relative positions of common metals in seawater. Please note that the second column follows the first.

in an electrolyte (seawater), zinc will be the anode and corrode and mild steel will be the cathode and not corrode. If mild steel (e.g. ship's hull) is connected to manganese bronze (ship's propeller), the mild steel now becomes the anode and corrodes and the manganese bronze propeller the cathode, which does not corrode.

This type of corrosion cell, consisting of two dissimilar metals, is easy to identify, but much more localized corrosion based on small size effects can be as effective in causing damage by the creation of pits. Localized corrosion occurs in metal owing to slight differences within the metal itself. A few of the causes of such corrosion are given below.

9.3.2   Concentration cell

Pitting may start as a localized anodic area due to the presence of a dissimilarity (e.g. oxide particle, carbide). Once started, the solution around the anodic and cathodic regions will have different concentrations and the process will accelerate because as well as differing materials there are different concentrations assisting corrosion (e.g. rusting of a nail in water, water droplet on the surface of a steel sheet – see Fig. 9.11). The droplet will have a higher concentration of oxygen dissolved on the outside than at the centre. The anodic area therefore forms at the centre.

This type of corrosion can occur in the splash zone of a structure that is protected for subsea corrosion unless another form of protection is applied, such as painting, above the water line.

9.3.3   Intergranular corrosion

All engineering materials used in the construction of platforms and pipelines, etc. are polycrystalline in structure; that is, the material is made up of very small grains, millimetres or smaller in size. These grains will not all be the same composition if the material is an alloy, but consist of grains of two or more different phases. Therefore, at the grain size level of the material, many forms of dissimilarities exist.

In the case of a two-phase metal like plain carbon steel, each phase acts as a different metal so that grains of one phase are anodic while those of the other are cathodic. Thus, corrosion cells are set up between them, as in Fig. 9.12 and 9.13, so that the phase that is the anode corrodes while the other phase acts as the cathode.

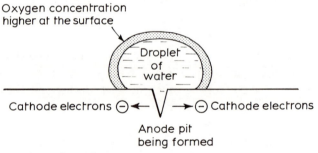

Fig. 9.11   Pit being formed.

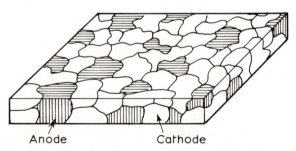

Fig. 9.12   Three-dimensional grain structure in a metal. Shaded grains are the α phase of the metal, unshaded are the β phase of the metal.

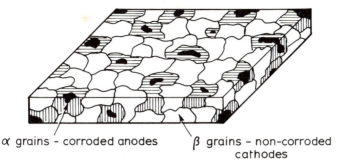

α grains - corroded anodes   β grains – non-corroded
                                          cathodes

Fig. 9.13   In the Galvanic Series, the α phase is above the β phase. The α phase
will therefore corrode.

Even in single phase alloys, preferential corrosion of one of the components
may occur; dezincification of brass is an example of this.

In the case of grain boundary corrosion, the higher energy regions of the
grain boundaries become the anodes and the main part of the grain the cathode.
This results in metal removal along the grain boundary in the form of a line (see
Fig. 9.14). Weld decay is an example of this type of corrosion, where the anode
forms along the weld toe.

Stress corrosion is a form of intergranular corrosion that increases in severity
when the material is subjected to a tensile load. The effect is to concentrate the
corrosion on a limited number of grain boundaries that are at right angles to the
direction of loading, as in Fig. 9.15.

This type of corrosion is seen in structural members supporting high tensile

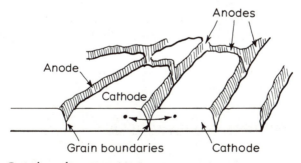

Fig. 9.14   Grain boundary corrosion.

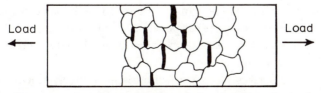

Fig. 9.15   Stress corrosion. Some grain boundaries at right angles to the load
direction are deeply corroded.

loads, and the rate of corrosion is much faster than that of a similar member subjected to a much lighter load.

The increase in energy that accelerates corrosion due to stress need not only be due to the application of a direct load. Stresses that are locked into a material (residual stresses) by manufacture or accidental damage may also cause this energy increase. Season cracking is the result of residual stresses and corrosion, and the rapidity with which the folds of the damaged part of a motor car wing rust compared with the surrounding unprotected material is an everyday example of the effect of locked-in stresses on corrosion.

Fretting corrosion occurs in certain situations where surfaces which are in contact move slightly; the corrosion of one or both of the surfaces can be very rapid. This could occur in the metal adjacent to clamps and collars if there is the slightest movement underneath them.

9.4   BIOLOGICAL CORROSION

Corrosion by marine biological action can be initiated in various ways:

(a) By the production of corrosive substances like hydrogen sulphide or ammonia, which result in direct chemical attack on the metal.
(b) By producing or actually being a catalyst in the corrosive action.
(c) By the reaction of sulphate-reducing organisms under anaerobic conditions. The most important of these are the bacteria *Sporovibrio desulfuricans.* These thrive in the reduced oxygen conditions created under heavy accumulations of marine growth, under thick deposits of corrosion products, or under mud. There are indications that the effect of this organism is to take the place of oxygen, which is unable to diffuse through the heavy marine growth, in the usual cathodic reaction.
(d) By the formation of concentration cells around and under the organisms.

9.5   CORROSION IN SEAWATER

Metals in seawater will be subjected to two main modes of attack: uniform attack and localized attack.

The former is where all the surface is corroded to the same degree. In this case, the useful life of a structure can be estimated and its progress monitored by regular thickness measurements. Unexpected failure should not occur if regular inspection is carried out.

Localized attack takes place in small areas in relatively slowly corroding structures, e.g. pitting or intergranular corrosion. The effect of localized corrosion is to lose the local integrity of the material by causing leakage, and to provide sites from which fatigue cracks (corrosion fatigue) or brittle fracture cracks propagate. These are much more difficult to locate and require much more meticulous inspection.

The rate at which corrosion takes place will depend on many environmental factors, such as the following.

9.5.1   Temperature

Most reactions are speeded up by an increase in temperature, by temperature cycling and by temperature differences, so that hot risers, exhausts and cooling water exits are sites that can corrode more rapidly than the rest of the structure.

Studies undertaken by the Dow Chemical Company showed that the corrosion rate of mild steel and a low alloy steel, in brines at a pH of 7.4, approximately doubled as the temperature was increased from 180°F to 250°F. Therefore, components like cooling water outlets and hot risers are particularly susceptible to corrosion and must be inspected regularly.

The effect of seawater temperature is illustrated in Fig. 9.16.

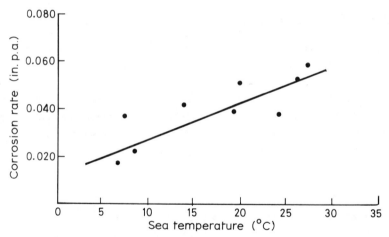

Fig. 9.16   The effect of temperature on corrosion of steel in seawater.

9.5.2   Water flow rate

The effect on the corrosion of a metal caused by increasing the flow rate is generally that the rate at which metal is removed is increased. If impingement of the flow on the metal or aeration takes place in the region of the surface, then a very much larger rate of metal removal is experienced locally. Pitting of ship propellers and pump and dredger impellers are good examples of this.

Tests carried out by P. Ffield show the straightforward effect of increasing velocity on the corrosion of steel pipes carrying seawater at different velocities (see Fig. 9.17).

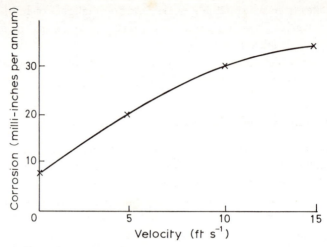

Fig. 9.17   Effect of seawater velocity on corrosion of steel at ambient temperature, exposed 38 days.

9.5.3   pH value of the water

The pH value affects the corrosion rate of metals. The modern definition of the pH value is the amount of hydrogen activity. This more recent definition replaces the old definition, which was that the pH value was the hydrogen ion concentration. The definition allows a number to be given to the pH value of any solution forming part of the electrochemical corrosion cell. This number basically measures the relative acidity or alkalinity of the solution with respect to pure water, which has a pH value of 7. Steel corrodes least when in a solution whose pH value is 11 to 12.

9.6   CORROSION PROTECTION

9.6.1   The Pourbaix diagram

Before we consider methods for the prevention of corrosion, we will briefly examine the way in which the corrosion process is influenced by the two main variables: the electrode potential and the pH value of the solution. These data are often presented in diagrammatic form, known as a Pourbaix diagram (see Fig. 9.18).

These diagrams are obtained from laboratory tests carried out under controlled conditions of constant temperature and no flow. It will be seen from Fig. 9.18 that there are three distinct possible states of corrosion, depending on electrode potentials and pH values:

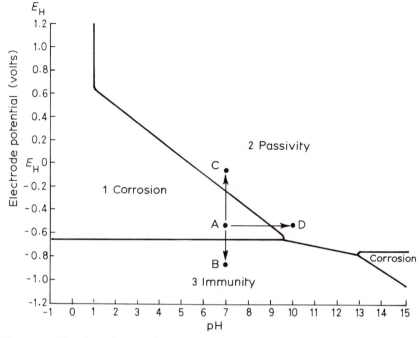

Fig. 9.18   Pourbaix diagram for iron in water.

(a) Corrosion.
(b) Passivity.
(c) Immunity.

(a)   Corrosion

At intermediate electrode potentials, corrosion takes place and metal is removed.

(b)   Passivity

At higher electrode potentials, we pass into the passivity region. This is the region in which a corrosion product film is formed, which in most cases is an oxide film. It is worth noting that the diagram only indicates that an oxide film is formed; it does not mean that the oxide film gives protection. The properties of the film must be known in order to determine this.

(c)   Immunity

At low electrode potentials, the rate of corrosion is so low that the metal is said to be immune.

9.6.2 Cathodic protection

If we consider a structure or component in an environment indicated by a point A in Fig. 9.18, we can see that corrosion will take place. A cathodic protection system would reduce the potential into the immunity region; that is, from A to B. The potential would therefore have to be greater than −0.68 V.

There are two ways in which this can be achieved. The first is to make the structure or component the cathode of an electrochemical cell by attaching to it a metal which is anodic to it. This type of protection is referred to as sacrificial anode protection, because the anode corrodes, thus providing the lowering of potential for the protected structure (see Fig. 9.19 and compare it with Fig. 9.7). A glance at Fig. 9.10 will show that magnesium, zinc and aluminium, and their alloys, are anodic to steel. Therefore, an anode made from one of these and attached to a steel structure would give protection to the structure. Pipelines, ships' hulls and boilers are often protected using this method.

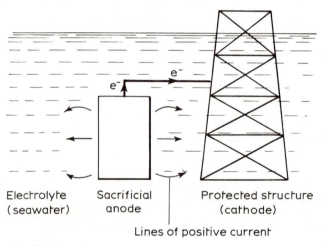

Fig. 9.19 Sacrificial anode protection.

Visual inspection of the sacrificial anode installation is required to see how much of the anode has been used up and to ensure that the electrical connection between the anode and the component being protected is intact.

The second method used to reduce the potential drop into the immunity region is to make the structure cathodic by impressing a current on it (see Fig. 9.20). This is often referred to as impressed current cathodic protection. With this system, the anode electrode can be a single metal anode electrode placed outside the structure or a series of smaller anode electrodes placed at certain points within the structure. The anode electrode used in an impressed current system can be expendable like ordinary steel, but these would of course require replacements, so it is much more usual to use permanent impressed

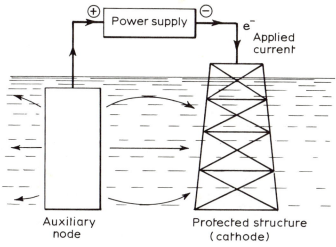

Fig. 9.20 Impressed current protection.

current electrodes that are only very slowly destroyed by the passage of the protective current. Electrodes of platinum-sheathed titanium, tantalum, niobium, lead silver alloys, silicon iron alloys and graphite can be used. The most common anode material in North Sea operation is platinum.

Visual inspection for this system of protection is required to determine any deterioration of the auxiliary anode metal and the integrity of the supply cable from the power supply to the anode. Any breaks or cracks in the casing of this cable will allow the supply cable at that point to act as another anode and as this cable is normally copper, the copper will soon be eaten away by the anode reaction, thus cutting off the current to the auxiliary anode.

One factor which merits consideration with this type of corrosion protection system is safety. The anode electrodes are often energized at up to 25 V and 1000 A (that is, about 100 times the voltage of a sacrificial anode system). If the anode electrode is located remote from the structure on the seabed, a relatively small number of these can be used and they can be located at a considerable distance from the steelwork in order to achieve a uniform current distribution.

The hazards to divers are all associated with the high intensity impressed current anodes and with damage to the supply cable insulation. As would be expected, most of the voltage drop is in the vicinity of the anode electrode and is reduced with distance from it, following an approximate logarithmic law. See Fig. 9.21.

Examination of the voltage drop curves through seawater indicates that approaching power impressed anode electrodes closer than 12 m (40 ft) could mean moving into a hazardous area. There is no hazard to divers from sacrificial anode systems and there is no hazard to divers on the structure from remote anode electrodes, provided, as already stated, there is no damage to the

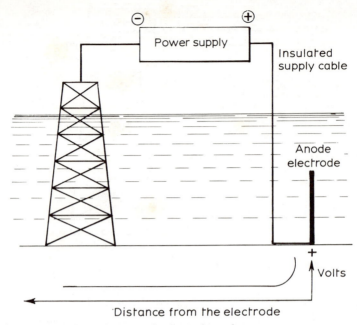

Fig. 9.21   Voltage drop from anode electrode to the structure.

electrode supply cable, which probably runs to the seabed via the structure. However, the requirement to turn off the impressed current protection system will be determined by the maintenance schedule.

If, however, the anode electrodes are mounted inside the steel structure, then it is impossible to know the form of the electric fields and the cathodic protection should be switched off before any diver enters the sea in the vicinity of the energized anode electrodes.

9.6.3   Anodic protection

If we consider Fig. 9.18, anodic protection is achieved by changing the operating conditions from A into the passivity region at C. This method is not widely used as there are two possible difficulties. First of all, if the potential is not raised sufficiently, then instead of the structure being at condition C it might still be in the corrosion region and all that has happened is an increase in the corrosion rate. Secondly, even if point C is reached, it only tells us that an oxide film is formed, not whether that oxide film will protect the metal.

9.6.4   Control of the pH value

If we again consider Fig. 9.18, we can see that there is a third way in which we can move the material out of the corrosion region. This is by changing the pH value of the electrolyte and moving the material to D. For example, steel corrodes least in a solution whose pH value is 11 to 12.

This system can be used as a method of corrosion protection in enclosed spaces. The classic case is the addition of caustic soda to boiler feed water; also, single point moorings (SBMs), when not in use, are filled with water containing an inhibitor to give controlled pH value for corrosion protection. The subsea wellhead chamber designed for a neutrabaric system of maintenance is another case in which control of the pH value of the water is a possible method of corrosion control.

The other piece of information that comes from the understanding that changes in pH value affect corrosion rates is that when equipment and rigs are moved from one location to another the pH value may well be altered. An acceptable level of corrosion activity at one site might therefore change on moving from one location to another (e.g. from the open sea to the outfall from a river estuary).

9.6.5   Protective coatings

Above the water line, the cathodic protection discussed earlier would not be effective, and localized corrosion would go on under isolated droplets of water. The normal method of protection used here is therefore by the application of additional surface layers, depending on the conditions.

(a)   Organic materials

Organic materials, such as oils and greases, give temporary protection. Paints, varnishes, plastic coatings or lacquers are more permanent, depending on the resistance of the surface to scratching, wear or erosion.

Additionally, there is now a peelable protective coating which may be applied underwater and removed easily when necessary, revealing a clean surface. This coating is primarily designed to protect the surface from marine fouling, but can also be used to clean areas, such as critical welds, wall thickness test points, and risers prior to inspection.

Trials have taken place to investigate the effect of this peelable coating on cathodic protection as well as to see if water trapped behind the coating during application would result in a corrosion cell. These trials revealed no corrosion behind the coating, but the coating is not yet used extensively in the field.

(b)   Inorganic materials

These include vitreous enamels, which are not generally seen in the marine environment but which are invaluable for higher temperature corrosion protection.

(c)   Metallic coatings

There are two classifications for these. The first is where the protected metal is isolated from the corrosive environment by a metal lower in the Galvanic Series (e.g. chromium, nickel and tin on steel). In this case, if the coating is penetrated, the metal being protected becomes the anode and corrodes away. An example of this in the offshore environment is the Monel wrapping encasing riser pipes on the production platforms. The second classification is where the protected metal is coated with a metal higher in the Galvanic Series. This is called galvanic protection, which is the same as cathodic protection described earlier (e.g. cadmium and zinc on steel).

9.7   CORROSION PROTECTION MONITORS

The amount of current from sacrificial anodes or from an impressed current system required for protection varies:

(a) From metal to metal.
(b) With the geometry of the structure.
(c) With differences in the seawater environment (temperatures, pH value, etc.).
(d) With any other factor that affects the resistance of the circuit.

Since the amount of current required for protection of any structure cannot be accurately predicted or distributed evenly through the structure, the method of checking for adequate protection is to measure the potential of the structure at various places on it.

This potential is measured by using reference electrodes, which are often called half cells (see Fig. 9.22). In seawater, there are two systems for doing this,

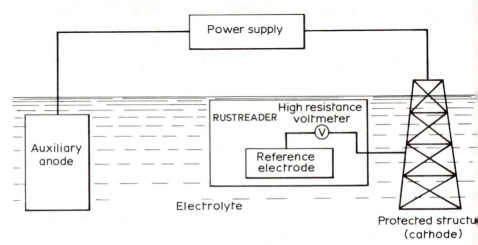

Fig. 9.22   Rustreader measuring electrode potential on a structure.

one being to use high purity zinc electrodes and the other to use a silver/silver chloride (Ag/AgCl) electrode or half cell. Two examples of instruments which use this latter method are the Roxby Bathycorrometer and the Morgan Berkley Rustreader. Both measure the electrode potential that exists at the point of contact made by the instruments. They are handheld instruments used by a diver but variations exist and are used by remotely operated vehicles (ROVs) and remotely controlled vehicles (RCVs). In all cases, the potential readings given by the instruments indicate whether or not the structure is protected at that point. It does not give a measure of the rate at which a structure is rusting, or the loss of material due to corrosion; the latter can be obtained by thickness measurement. An alternative method to these contact readings is to install permanent electrodes remotely around the structure. In this case, readings are displayed on a console in a control room.

When using the Bathycorrometer or Morgan Berkley Rustreader, the instrument should be soaked in seawater for about an hour before use. This is to establish the chlorine ion concentration in the cell with seawater via a porous membrane. Care must be taken not to cover the inlet in the instrument body to this membrane when the instrument is being used.

To calibrate the instrument, a zinc block is usually supplied and when this is used a voltage of −1 V is measured. The reading on the dial or digital display does not show this minus sign.

When the instrument is used on a piece of unprotected steel, a reading of 0.65 V is shown, but if the steel is protected, a reading of 0.8 V is shown. Again, the minus sign is omitted from the display, but from the Pourbaix diagram (Fig. 9.18) we can see that the potential must be more negative for protection to occur.

Inspection requirements for corrosion are therefore as follows:

(a) Visual inspection of anode (both sacrificial and auxiliary) for wear.
(b) Visual inspection of the electrical connection of the sacrificial system to see that it is intact and of the impressed current system to see that there are no breaks in the casing of the supply cable.
(c) Potential measurements on the structure to see that it is still the cathode of the system.
(d) Corrosion damage by visual inspection and thickness measurement by ultrasonic thickness meter.

10 Magnetic particle inspection (MPI)

10.1 HISTORY OF MAGNETISM

Even in very early times, it was known that magnetite (an iron ore) attracted small pieces of iron. Also, if magnetite were suspended, it would rotate and align its longest axis in a north–south direction. This gave rise to its name 'lodestone', derived from an Anglo-Saxon word meaning 'way' or 'course'. The directive property of magnetite was utilized in early navigational devices.

The first method of forming an artificial magnet was discovered by the Frenchman de Magnette in 1600. He found that heating and hammering an iron bar produced a 'power' in the bar that enabled it to attract pieces of iron. This power, named after him, was called magnetism.

In 1819, the Dane Oersted observed a relationship between electricity and magnetism. He noticed that when a compass was placed near a current-carrying wire, the compass needle showed a deflection. This phenomenon is now known as electromagnetism.

10.2 TYPES OF MAGNETISM

There are three types of magnetism (see Fig. 10.1).

10.2.1 Ferromagnetism

This is shown by materials which can be strongly magnetized and which show good magnetic properties.

10.2.2 Paramagnetism

This is shown by materials which are weakly attracted by strong magnetic forces.

Ferromagnetic	Paramagnetic	Diamagnetic
Iron	Platinum	Bismuth
Nickel	Palladium	Antimony
Cobalt	Most metals	Most non-metals
Steel	Oxygen	Concrete

Fig. 10.1 Magnetic types exhibited by certain materials.

10.2.3 Diamagnetism

This type of magnetism is shown by materials that are repelled by a strong magnetic field. This externally applied magnetic field induces a 'like' magnetic field within the material; hence, repulsion occurs.

10.3 THEORY OF MAGNETISM

In ferromagnetic materials, the atoms are gathered together in groups called domains. These domains have a magnetic moment, one end acting as north pole, the other as south pole. This magnetic moment is created by the combined effort of the motion of electrons around the nucleus of the atom and by electron 'spin', which is the rotation of the electron about its own axis.

When the material is unmagnetized, the domains lie distributed randomly and their magnetic effects cancel each other (see Fig. 10.2).

If an exterior magnetic field is applied to the material, the domains are aligned north to south in a common direction (see Fig. 10.3). Hence, one end of the material will be the north pole and the other the south pole.

10.3.1 Polarity

When the material is magnetized it has a north and a south pole. These poles are located at opposite ends of the material and magnetism seems to be

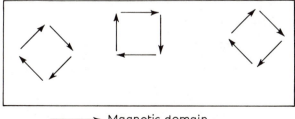

———➤ Magnetic domain

Fig. 10.2 Material in an unmagnetized state.

South pole   North pole

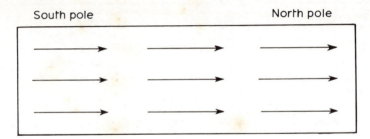

Fig. 10.3   Material in a magnetized state.

concentrated at these points. The 'north' or 'north seeking' pole of a magnet is said to be the pole that points towards the earth's north pole, the south pole of the magnet pointing towards the earth's south pole. Magnetic poles show attraction and repulsion, like poles repelling (see Fig. 10.4) and unlike poles attracting (see Fig. 10.5).

10.3.2   Magnetic field

This is described as the area surrounding the magnet in which the magnetic forces exist. The magnetic field is represented by lines of force or lines of magnetic flux. These lines are purely imaginary and were introduced by Michael Faraday as a means of visualizing the distribution and density (flux density) of a magnetic field. Magnetic flux is measured in webers and the symbol used to indicate magnetic flux is ϕ (phi).

These lines can be visualized as travelling from the north to the south pole externally (see Figs 10.4 and 10.5) and from the south to the north pole

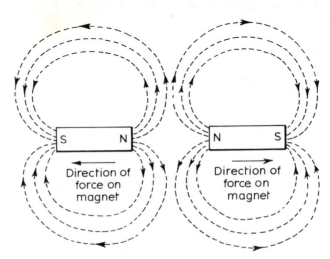

Fig. 10.4   North pole repels north pole.

→ Direction of force on magnet

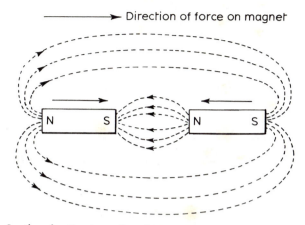

Fig. 10.5   South pole attracts north pole.

internally, to form a continuous closed loop (see Figs 10.6 and 10.7). They do not cross and they seek the path of least resistance. They are like stretched elastic cords, always trying to shorten themselves. A magnetic field exists within and around a magnet and around a current-carrying conductor. Lines of magnetic flux which are parallel and in the same direction repel one another (see Fig. 10.4).

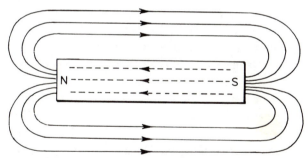

Fig. 10.6   Lines of force on a bar magnet.

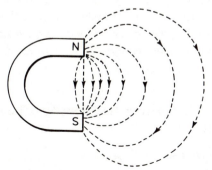

Fig. 10.7   Lines of force on a horseshoe magnet.

The magnetic field produced by a current-carrying conductor forms closed circles perpendicular to the conductor (see Fig. 10.8). The direction of the magnetic field is given by a right hand rule: the thumb points in the direction of the current and the fingers in the direction of the magnetic field.

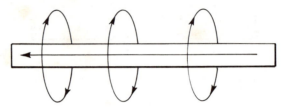

Fig. 10.8   Lines of force on a current-carrying conductor.

10.3.3   Flux density

Magnetic flux density is a term used to describe the strength of magnetism within a specimen. When a magnetizing force is applied, lines of force of magnetism will be created within the specimen, the number of which depends on the size of the magnetizing force. These lines represent the magnetic force, and the flux density is the term applied to the quantity of them that emerge per unit cross-section of the specimen. In the case of a permanent magnet, flux density is measured in tesla (1 tesla $= 1$ weber per m$^2 = 10^4$ lines per cm^2). The symbol for flux density is B. In the case of an electrically induced magnetic field, the unit of flux density is defined as the density of a magnetic field in which a conductor carrying 1 ampere at right angles to that field has a force of 1 newton per metre acting on it. The unit is still the tesla.

When a magnetic field is induced in a workpiece by an electric current flowing in a cable arranged either as a coil or parallel conductors, the imput electrical force is called the magnetizing force H, and can be defined in the following way. Let us consider a very long coil (like a solenoid) uniformly wound with N turns per metre and let the current flowing in this coil be 1 ampere. Then the magnetizing force H in this coil is $N \times 1$ ampere-turns per metre. (The latest practice is to omit the word 'turns' so that the units of H are amperes per metre (A m^{-1}).) This equation for the magnetizing force is the basis of the calculations for how much current to use to set up an adequate magnetic field for Magnetic Particle Inspection.

10.3.4   Relationship between flux density (B) and magnetizing force (H)

In free space, this relationship is $B = \mu_0 H$.

$B =$ flux density

μ_0 = permeability (this is the ease with which a material accepts magnetic flux) of space

H = magnetizing force.

In a magnetic material, however, B and H can vary independently and the coefficient of B/H is called the relative permeability and is often given the symbol μ (so that for air $\mu = 1$, but for certain nickel iron alloys it can be as high as 100 000). This is not a constant value, but varies with B and H. The value of relative permeability for various levels of magnetizing force H is given as graph B for steel in Fig. 10.9.

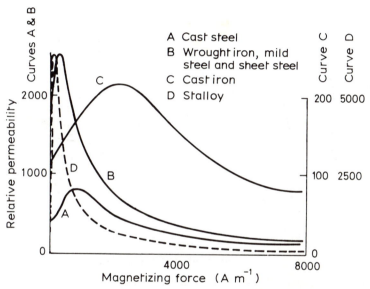

Fig. 10.9   Variation of relative permeability with magnetizing force.

When magnetizing a piece of steel using an electrical magnetizing force, we do not get a uniform increase in flux density with a uniform increase in magnetizing force (see Fig. 10.10). The flux density rises rapidly to point A on the graph and thereafter rises very much less rapidly as the condition approaches saturation of the magnetic field; i.e. there is a very small rise in flux density for a very large rise in magnetizing force.

The shape of this characteristic and the saturation value varies from material to material.

Figure 10.11, which shows flux density values for different values of magnetizing current, is for a field being set up in one direction by a DC supply. If the coil uses an AC supply, then the magnetic field will experience a complete reversal of direction for each cycle and successive cycles will trace out a loop for the relationship between magnetizing force and flux density. This is known as the 'hysteresis loop' (see Fig. 10.12).

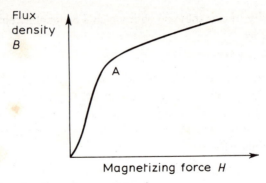

Fig. 10.10 Flux density and magnetizing force.

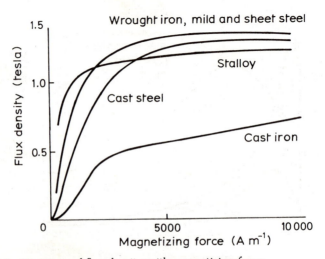

Fig. 10.11 Variation of flux density with magnetizing force.

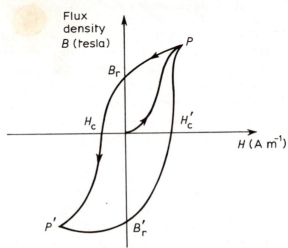

Fig. 10.12 Hysteresis loop.

OP is the initial magnetization curve as seen in Fig. 10.12. After P, the current reduces and the flux density reduces. The curve then passes through B_r to H_c on to P', the maximum flux density for the magnetic field in the opposite direction. The current now reduces again and the flux density reduces to B'_r and as the current increases in the opposite direction the curve climbs to point P at maximum H. Successive cycles trace out the hysteresis loop P, H_c, P', H'_c, P. The loop tells us that energy is used up during each cycle and is dissipated in the form of heat and while this heating effect is rapidly dissipated under water, on land the material will get quite hot. The value B_r indicates that when the magnetizing force is removed the material will still be magnetic. The name for this is 'remanence' (or 'residual magnetism' in the USA) and H_c is the magnetizing force required to remove it. This is known as 'coercivity'. In practice, the residual magnetism is removed by gradually reducing the magnetizing force (lowering the current) so that the loops get smaller, thus reducing the residual magnetism.

The terms 'permeability' and 'retentivity' still remain in use although the ideas of relative permeability and remanence have given them a more fundamental explanation. In this older explanation, permeability was defined as the ease with which a material could be magnetized, and of course it is only some of the ferromagnetic group that are usefully magnetized. Those that are easy to magnetize are soft iron and low carbon steels and these were said to have high permeability, while high carbon steel, which required more magnetizing force to produce the same amount of magnetization, was said to have low permeability. When the magnetizing force is removed, the amount of magnetism retained varies from material to material and this effect was known as 'retentivity'. The amount was said to be related to the permeability of the material and the general rule was that materials with high permeability have low retentivity and those with low permeability have high retentivity. Modern permanent magnets are generally made from low permeability/high retentivity special alloys which have been subjected to large magnetizing forces. Another term associated with permeability is reluctance, which is used to define the resistance of any particular material to the flow of a magnetic field.

10.4   MAGNETIC PARTICLE INSPECTION

This system of inspection is based on the phenomenon that the path of the magnetic flux in a ferromagnetic material is distorted because inhomogeneities (such as cracks, blowholes, inclusions, grain boundaries, etc.) have different magnetic properties to a greater or lesser degree than the surrounding material. All systems of magnetic non-destructive testing need some method of detecting this distortion of the magnetic flux, often called 'leakage flux'.

10.4.1   Magnetization

The magnetic field can be set up in a magnetic material in the following ways:

(a) Induced magnetism, by the use of a permanent magnet.
(b) Induced magnetism, by passing an electric current directly through the workpiece.
(c) Induced magnetism, by passing an electric current through a conductor close to the workpiece.

(a)   Use of permanent magnets

This also applies to electromagnets. A U-shaped magnet is used, the workpiece completing the magnet path or circuit between the poles. The lines of flux that normally exist between the poles (see Fig. 10.7) are concentrated in the workpiece, instead of returning through the air, as the workpiece completes the magnetic circuit (see Fig. 10.13). The direction of the magnetic field set up using a permanent magnet is shown in Fig. 10.14. The maximum disturbance to the magnetic field and hence the maximum flux leakage are caused by defects that are at right angles to the field. This of course is true no matter how the magnetic field is produced.

In BS 6072: 1981, the suggested strength for a permanent magnet is a force of 18 kg (40 lb). If the magnet is too weak, then the field will be insufficient to give a clear magnetic pattern; if it is too strong, then a dense accumulation of particles will make the patterns difficult to interpret, especially in the region of the pole pieces.

On straight workpieces such as plate and cylinders, good contact between the pole pieces and workpieces is easily obtained by having shaped pole pieces, for example, flat for a flat plate (see Fig. 10.15) and radiused for cylindrical-shaped workpieces (see Fig. 10.16).

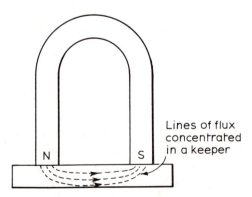

Fig. 10.13   The keeper completes the magnetic circuit concentrating the lines of flux in the keeper.

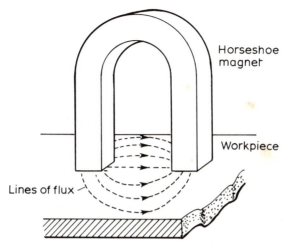

Fig. 10.14 Lines of magnetic flux in the workpiece caused by a permanent magnet.

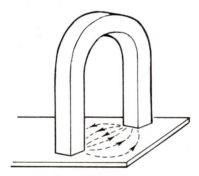

Fig. 10.15 Flat pole pieces for plates.

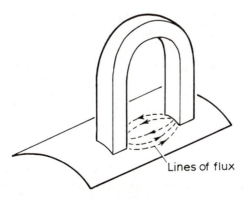

Fig. 10.16 Radiused pole pieces for cylinders.

For more complicated shapes, for example when examining the weld at the joint in a node, the two arms of the magnet need to be joined and the pole pieces need to roate, as well as being shaped so that a good contact can be made with the workpiece (see Fig. 10.17).

Electromagnets have the advantage that they can be applied to and removed from the workpiece easily when the current is off, but this system does need an electrical supply. Texas Magnetics of Houston, Texas, manufacture a magnetic inspection unit (WL 1UW) based on their land-based model WL-1. The unit is powered by a 12-V battery and the electromagnet gives about 18 kg (40 lb) of lifting force. Parker Research Corporation of Dunedin, Florida, also produce a magnetic inspection unit. The unit requires 8 A at 115 V AC, but has the capability of working in AC or DC mode. The probe has movable arms to adjust to different contours. Oilfield Inspection Services of Great Yarmouth, Norfolk, also produce an electromagnet.

(b)   Induced magnetism by passing an electric current directly through the workpiece or section of the workpiece

This is achieved by using contacts or prods. Sometimes these are clamped on and at other times they are hand held. This method of magnetization is recommended in section 14.4 of Det norske Veritas, Rules for the Design, Construction and Inspection of Offshore Structures, 1977, Appendix 1, In-Service Inspection.

Passing a current through the workpiece produces a circular field between the contact points (see Fig. 10.18). The magnitude of the current used will depend on the plate thickness and prod spacing, but will be of the order of 100 A per 25 mm (0.98 in) of separation for flat plate and 150 A per 25 mm (0.98 in) of separation for circular sections.

When using this technique, care must be taken to ensure that the contact areas are sufficiently clean to pass the current required (high amperage, often in excess of 1000) without arcing or burning. (If low voltage is used, this will assist in preventing burning.) The magnetizing current should not be turned on until after the prods have been positioned on the surface, and should be switched off before they are removed. The prods can be separate so that it requires one diver to position and hold the prods. This gives complete flexibility of positioning although it requires two divers to carry out a magnetic particle inspection. Alternatively, the two prods can be mounted as the two arms of a U-shaped probe, and this is the usual application, as it allows one diver to perform the task.

Direct or alternating current can be used for magnetizing the workpiece. Direct current produces fields which penetrate deeper into the metal than alternating currents, which because of the 'skin effect' are confined to the surface

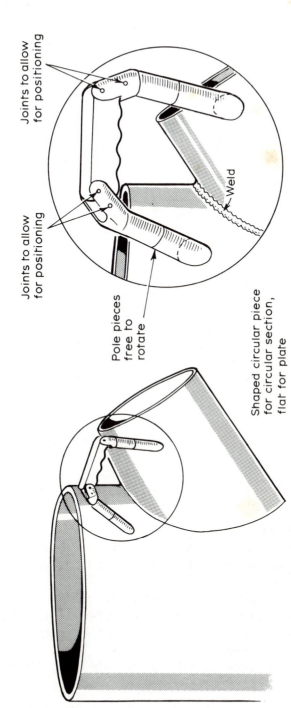

Joints to allow for positioning

Joints to allow for positioning

Weld

Pole pieces free to rotate

Shaped circular piece for circular section, flat for plate

Fig. 10.17 Magnet with adjustable pole pieces.

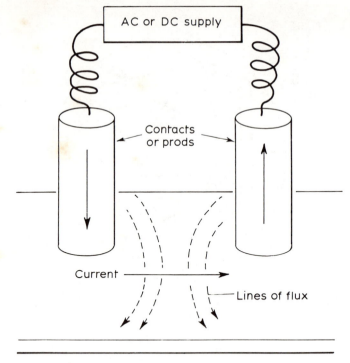

Fig. 10.18   Lines of magnetic flux for prods passing current through the workpiece.

of the metal. Direct current is recommended in water because it is safer for the operator.

(c)   Induced magnetism using a coil

As mentioned earlier, a conductor carrying a current induces a circular magnetic field around it. Therefore, ferromagnetic materials near the conductor will be in this magnetic field and lines of magnetic flux will be concentrated in them as they have less magnetic resistance than air.

A differently shaped magnetic field will be produced if we wind the conductor in the form of a coil around the material to be tested (see Fig. 10.19). A piece of ferromagnetic material placed in the coil, or outside parallel to the coil, will experience longitudinal magnetism.

The procedure for this method is laid out in BS 6072 which recommends that:

(i) The area under inspection should be within the coils and when the coils are moved it is in coil-length intervals.
(ii) The peak value for the direct current should have a minimum value of:

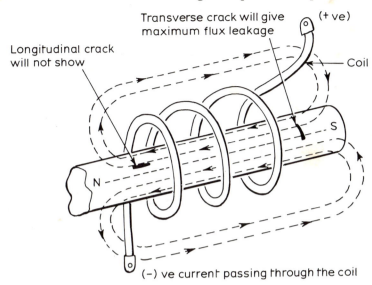

Fig. 10.19   Longitudinal magnetization.

$$I = 7.5 \left(T + \frac{Y^2}{4T} \right)$$

where I is the peak current (DC); 7.5 is a constant (based on the permeability of the material); T is the wall thickness (in mm); and Y is the spacing (in mm) between adjacent windings in the coil.

It is common practice to use three to five turns in the coil.

A magnetic field that is transverse can be produced by looping the conductor (see Fig. 10.20). This produces a transverse field between the two sides of the loop. The loop has to be positioned so that the current in the two sides of the loop is moving in the same direction or else the magnetic fields will tend to cancel instead of reinforce.

This method of magnetizing ferromagnetic materials offers development potential for producing adequate magnetic field for inspecting structures that have complicated shapes: for example, nodes and intersections.

It has been shown by Lumb and Winship in their paper on Magnetic Particle Crack Detection printed in *Metal Construction*, July and August 1977, that this method of magnetizing can give good uniformity in the magnetic field between the parallel sides of the loop.

BS 6072 recommends that:

(i) The test site must be in close proximity to the current flowing in one direction. This means that return cables must be as far as possible from the test site, and at least 10*d* away where *d* is the width of the test site.

(ii) The cable should be moved at intervals of not more than 2*d* (*d* as in (i)

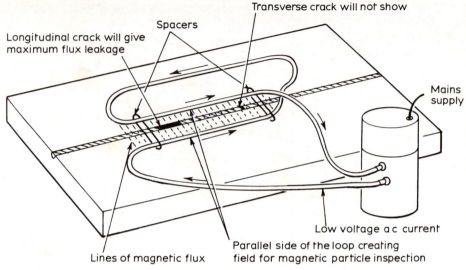

Fig. 10.20   Transverse field produced by looping a coil.

above). (This ensures that complete coverage of the workpiece is achieved as the energizing cable is moved successively across it.)

(iii) The cables must be insulated.

(iv) The width of the inspection area and the current are related by the formula:

$$I = 30d$$

where I is the peak current (DC); 30 is a constant; d is the width of the inspection area in mm.

10.4.2   Types of MPI equipment

The application of these principles leads us to examine the types of equipment offered by manufacturers. As we are dealing with underwater inspection and as there is no intention of giving a complete catalogue of this type of equipment, let us confine the list to only those types of equipment with an underwater application.

Prior to 1979, the equipment available in the UK was manufactured largely by Ardrox and Magniflux, with Osel being in the picture. Ardrox and Magniflux produced and still do produce permanent magnets. During this period, Det norske Veritas (DnV) came up with an underwater prod system, which was about the first time any firm had tried to put current flow equipment under water. The DnV model was not widely adopted, however, probably for two reasons. The first is that although they do work, prods are very difficult to use under water, and the second is that the unit was very heavy and bulky. One adaptation of this unit which was seen in the UK was a unit made by Mapel which suffered from the same problem.

Around 1979, Osel produced their composite unit. This was loosely based on a land-based design and the unit was capable of being used to power flexible cables (which can be made into coils), prods, or an electromagnet. This unit was designed and marketed jointly with OIS (Oilfield Inspection Services) and proved quite popular in its first season. It was much smaller than the DnV model and very much less bulky. The cables were also compatible for use on the circular members to be found in the North Sea. This joint design has provided the basis for three other units and two of them at least are in current production. Bix produced a unit using flexible cables and with the ability to be used to power prods, but it was of fixed output and had more bulk than the Osel/OIS version. Osel and OIS have split up since their joint venture, and now produce new versions of the original design. The new OIS unit is the smallest and the lightest on the market at the moment. The new Osel unit is based on the same lines as the first one with various improvements incorporated into the design in the light of field experience. Both the OIS and the Osel units are in current production.

The latest designs seen in the North Sea are by a German group called MMA and by Thorn. A computer is used to interpret the magnetic tape that the MMA current flow equipment produces. The Thorn equipment is a new departure and is the first underwater adaptation of an eddy current technique. This equipment looks very promising and may prove to be a market leader of the future. It is dealt with in more detail in Chapter 14.

10.4.3   Continuous and residual magnetization techniques

Both types are used for land-based operations, but continuous magnetization is mainly used for underwater inspection. Using this technique, the magnetizing of the workpiece is carried out at the same time as the inspection medium (e.g. magnetic particles) is being applied and examined.

The residual technique uses only the magnetism that is left in the material once the 'magnetizing force' has been switched off, to attract, hold and orient the magnetic particles.

The continuous technique is generally regarded as being more sensitive, but indications of defects other than mechanical defects such as cracks and holes can be given; for example, leakage fields from the coil wrapped around a workpiece, and flux leakage in the vicinity of a prod or pole from defects such as grain boundaries due to high flux density in these regions. The closer the magnetic field is to saturation the more sensitive it is to flux leakage due to small order inhomogeneities in the material (see Fig. 10.21). BS 6072, Method for Magnetic Particle Flaw Detection, suggests a minimum flux density of 0.72 T.

Residual magnetic fields are much weaker than those produced by the

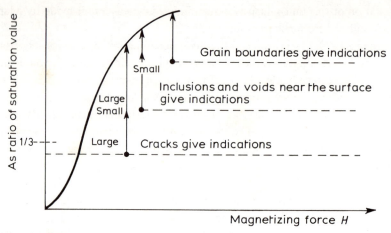

Fig. 10.21   Variation of types of defect giving indications as the field of strength increases.

continuous technique and therefore have a lower sensitivity, but of course this technique does eliminate false indications.

10.4.4   Demagnetizing

After a workpiece is magnetized for inspection, a residual magnetic field will remain, and the strength of this field will depend on the retentivity of the material. This should be removed if the inspection schedule calls for it, because the presence of the residual magnetic field is likely to disturb equipment such as guidance systems or electronic instrumentation.

10.5   DETECTION OF THE MAGNETIC FIELD

In underwater inspection, it is essential for the inspector to know that a magnetic field does actually exist between the prods or the coils where he thinks the field has been set up. This could be achieved by providing the inspector with a meter. It would need to be an underwater version of the land-based gaussmeter, of which there are several types. However, at present, such meters are not available.

Two other detection devices are available: the Burmah–Castrol strip and the Berthold penetrameter. The Burmah–Castrol strip is a strip of steel in which there is a fine cut completely encased in a non-ferromagnetic material such as brass. When placed in the magnetic field and sprayed with a magnetic ink, the presence of the cut in the steel will be indicated only if the magnetic field is alive. Sets of these strips with varying degrees of sensitivity are available. The correct type for offshore applications indicates a correct field of 0.72 T.

The Berthold penetrameter is shown in Fig. 10.22. The maker's information indicates that when the penetrameter is placed on the workpiece being magnetized, some of the induced flux is 'shunted' to two slits cut at right angles in the soft iron cylindrical test element. The flux leakage at these air gaps in the test element can be detected by applying magnetic ink to the thin plate that covers the test element. The indication is a thin black line where the particles in the ink have collected at the flux leakage. The direction of the field can be deduced from which line of the cross is shown up by the ink. The field lies at right angles to this line. The sensitivity of the penetrameter to field strength can be controlled by varying the distance between the test element and the plate that covers the test element, and so can be calibrated with a gaussmeter, to be used as a 'go/no go' gauge for field strength. The control of distance between the test element and the cover plate is often achieved by the use of non-magnetic shims.

When not in contact with the workpiece, the penetrameter also gives an indication of the presence of the workpiece as the lines of flux in air will tend to concentrate in the soft iron element.

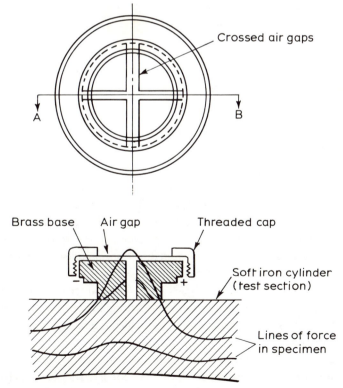

Fig. 10.22   Berthold penetrameter.

It is good practice to check that a magnetic field has been established in the area of the workpiece being inspected, and the use of an indicator such as a Burmah–Castrol strip can be easily incorporated into the test procedure without increasing the number of operations required of the diver-inspector. The strip, which is attached to the equipment, can be sprayed with ink and checked in the same operation as the workpiece inspection. It is strongly recommended that this check is carried out. BS 6072 bans the use of these instruments when using magnetic flow techniques.

10.6   DETECTING THE DISTORTION IN THE MAGNETIC FIELD (FLUX LEAKAGE)

The effectiveness of the magnetic inspection is determined by the sensitivity with which we can detect the change in the magnetic field due to the presence of a defect. There are three ways in which this can be done.

10.6.1   By moving a compass over the surface of the magnetized workpiece

With this method, the needle aligns itself with the magnetic field and indicates any local distortion of the field by a change in angle. This method is too slow and cumbersome for underwater use.

10.6.2   By using a search coil

This technique requires a search coil to scan the magnetized workpiece. The magnetic flux from the workpiece induces a voltage in the search coil as it moves over the surface of the workpiece. The voltage depends on the gap between the workpiece and the coil, the strength of the magnetic field, the speed with which the coil moves over the surface of the workpiece, and the relative angle of magnetic field to the scan direction. Flux distortion due to the presence of a flaw varies the induced voltage generated.

This method is used on land to inspect pipes. These are magnetized using an electric current and the search unit is propelled along the outside of the pipe. A flaw in the pipe causes a variation in the flux, which is indicated by a change in the induced voltage.

This method of detection has possibilities for underwater inspection, but in the first instance will probably be developed for fixed shape structures or possibly pipelines.

10.6.3   By visual observation of the distortion of the magnetic field as evidenced by the patterns caused by magnetic particles

In general, magnetic powders are available in dry powder form or as a liquid suspension. The surface of the workpiece should be thoroughly cleaned. On application, the magnetic powder is attracted to the defect, so that the defect is identified visually by a dense line of powder along its edge. The powders are finely divided ferromagnetic particles which have a high permeability and low retentivity. The particles may be coated to give greater mobility, coloured to give maximum contrast on the workpiece, or treated in such a way that they fluoresce when exposed to ultraviolet light.

10.7   APPLICATION OF MAGNETIC PARTICLES

10.7.1   Dry powder

This is obviously not used under water, except in hyperbaric chambers. The powders are often coloured and coated to increase mobility, and they are puffed onto the workpiece as a cloud. Sometimes, mobility is assisted by mechanically vibrating the workpiece or the particles are moved by varying the field, e.g. half wave rectified AC or pulsed DC.

10.7.2   Magnetic inks

Magnetic powders in liquid suspensions are referred to as magnetic inks. The liquid used for suspending the magnetic particles is either water or a light petroleum distillate. The specifications for the land-based inks and powders are given in BS 4069, Magnetic Flaw Detection Inks and Powders, and these specifications require that magnetic inks shall:

(a) Have uniform suspension of particles when agitated.
(b) Not contain anything that may be likely to cause discomfort to the users.
(c) Not corrode or damage in any other way the surface of the workpiece.

They may also contain small quantities of other ingredients within specified proportions.

10.7.3   Black (or non-fluorescent) inks

These inks are not used in water. After the surface has been thoroughly cleaned, it is covered with a thin layer of white (or some other contrasting colour) paint. The paint is applied manually or with an aerosol and must be of a type that is easily removed. Once the workpiece has been magnetized, the ink is painted or sprayed onto the surface of the workpiece, and the leakage flux from any defect attracts and traps the particles at the site of the defect. Because of the contrast between the colour of the particles and the background, the defect can be detected.

The BS specification for the composition of non-fluorescent ink is:

(a) Ferromagnetic particles shall not be less than 1.25% and not more than 3.5% by volume.
(b) Other solid constituents shall be not more than 10% by mass of the ferromagnetic content.
(c) Carrier fluid makes up the remainder.

It is recommended in BS 6072 that the level of the ambient light should be at least 500 lux (lx).

While this system is not used under water because of its dependence on ambient light, it should certainly be considered for use in offshore superstructure inspection above the waterline.

10.7.4   Fluorescent inks

These are solutions with a suspension of a ferromagnetic powder that fluoresces when exposed to ultraviolet light. The property of fluorescents is the ability to absorb light energy of an invisible wavelength (for which ultraviolet radiation is ideally suited) and re-emit this energy at the wavelength of visible light.

The BS requirements for fluorescent inks are:

(a) Ferromagnetic particles shall be not less than 0.1% and not more than 0.3% by volume.
(b) Other solid constituents shall not be more than 10% by mass of the ferromagnetic content.
(c) Carrier fluid makes up the remainder.

It is recommended in BS 6072 that the background light be less than 10 lux and the black light necessary to cause the ink to fluoresce shall be as per BS 4489. The minimum output for an ultraviolet light, as laid down in the British Standards, is 50 lux at the viewing distance.

During inspection, the workpiece is thoroughly cleaned and then magnetized using one of the methods described earlier. The fluorescent ink is then squirted across the surface to be inspected. This can be done by the diver squeezing a plastic bottle ink dispenser, or alternatively, the container with the ink can be located at the surface. In the latter case, the ink is fed by umbilical to the diver from a pressurized container. An inverted air cylinder makes a useful size container for the ink at the surface. A dispensing nozzle for the ink is attached to the ultraviolet (u/v) light housing and the ink supply is controlled by a trigger on the u/v light housing grip (see Fig. 10.23). This allows the diver to dispense the fluorescent ink and operate the u/v light with one piece of equipment in one hand. While the plastic bottle applicator is the simpler item of equipment, the surface supply system is simpler for the diver to operate.

The cloud of ink soon clears from the workpiece, leaving the ferromagnetic particles attracted to the surface with thicker deposits in the region of defects. When viewed with ultraviolet light (sometimes referred to as 'black light'), the particles fluoresce, giving areas of bright light where the particles are thicker. Where the ambient light intensity is high, as in the Arabian Gulf, these tests are carried out at night.

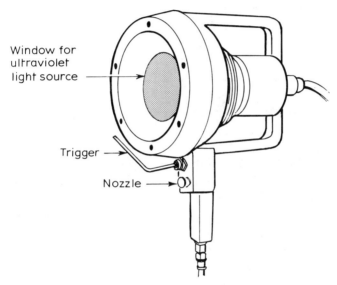

Window for
ultraviolet
light source

Trigger

Nozzle

Fig. 10.23   Osel ultraviolet light housing with magnetic particle injection system.

For land-based installations, the specifications for ultraviolet lights are given in BS 4489.

10.8   MPI SYSTEMS CURRENTLY IN USE

Currently, there are some eight systems that can be seen in operation. If the method of operation of the Osel/OIS composite unit is looked at in detail, this can be used as a model for all the other units and individual variations can be discussed.

The composite unit comprises a surface control unit, a multi-core conductor cable, and a submersible transformer unit incorporating an ink distribution system and an ultraviolet light.

The surface control unit incorporates electrical safety devices and isolating transformers because of the hazardous conditions in which it is operated, and controls to turn on the power to the current flow cables or prods, as well as to power the ultraviolet light. The amount of power available to the cables and prods is variable on the surface, and can be adjusted to suit the work in hand.

The submersible unit comprises terminals for cables or prods or electromagnets, whichever are selected. Terminals for the ultraviolet light and an ink reservoir which will hold 1 litre of ink concentrate and incorporates a paddle to keep it agitated, are also included.

The ink distribution system in this design consists of an electrical pump which draws in a metered amount of liquid from two sources. One source is an ink concentrate reservoir which contains a prepared mixture of ink ten times stronger than BSI recommendations. The other is a seawater intake. The metering allows one part of ink to nine parts of seawater to pass through the pump to a mixing chamber and then to the light. This elegant system causes considerable problems due to its sophistication and in later developments it has been designed out.

The above format is the basis on which the following units work:

DnV, Mapel, Bix, OIS and Osel.

The variations are only in the method of providing ink. With DnV, Mapel and Bix, the ink is made up to BSI requirements and a 10-litre batch is poured into a container incorporated into the submersible unit. This container is pressurized via its own air supply (usually a SCUBA bottle) and is fed under pressure to the ink distribution nozzle. Ink agitation is achieved by incorporating an electrically driven paddle into the reservoir. With a new version of the Osel unit, they also have reverted to a pressure system.

With the OIS unit, there is a new departure which works very well indeed. The ink is made up to BSI standard again, but this time it is poured into a 10-litre soft plastic bag, which is protected by a stainless steel frame and which incorporates a mechanical pump, which is driven electrically. Once connected, the ink is constantly agitated. Distribution of the ink is achieved by plumbing a tee-piece into the fixed pipe and providing a take-off via a Helle–Hanson connector for the light.

The MMA unit consists of a surface computer unit, a surface control unit, a submersible transformer unit and a flexible holder which is designed to accept 125 mm (4.9 in) wide magnetic tape – the same sort that is used in tape recorders, etc. The diver-inspector is provided with a quiver full of tape cut to the correct length and has to load these one at a time into the holder. The holder is then placed over the weld and when satisfied that it is correctly positioned, the diver-inspector depresses the on/off switch for about five seconds. While the current is flowing, a magnetic imprint representing the surface of the weld is recorded onto the tape. The diver must then ensure that the exposed tape is not subjected to any further magnetic fields – hence, the quiver. As soon as the tape is brought to the surface, it is dried off and then fed into the computer. The computer scans the tape and then prints out on paper a profile of the weld cap, which then has to be interpreted.

10.9   RECORDING

After visual observation, the inspector needs to note and report the position of the defect, type of defect, size, shape and orientation of the defect.

Currently, it is possible to photograph the workpiece under water with a still camera and flash. Alternatively, once the crack is found, it could be retested with Miglow magnetic ink and rephotographed using white light. Miglow is the trade name for fluorescent ink which fluoresces in daylight.

On land, the magnetic pattern is taken as a record either by a pressure-sensitive tape pressed onto the area so that the pattern is recorded by the magnetic particles adhering to the tape, or by having the particles in a colloidal suspension which solidifies, thus fixing the pattern. A pressure-sensitive tape system is being developed and used in underwater inspection. It consists of a magnetic tape which is exposed to a magnetic field while being pressed onto the surface of the workpiece. The tape is then brought to the surface and the magnetic patterns are read and interpreted electronically, as outlined above in the MMA method.

With the latest development of a moulding system by BP Chemicals Ltd, it is possible to take a cast and thus reproduce a three-dimensional record of any defects found by MPI. This method is marketed as Aquaprint, and is outlined in Chapter 8.

10.10   MAGNETIC PARTICLE TEST PROCEDURE

A typical magnetic particle test procedure would be as follows:

(a) The specification for the test is carefully assessed.
(b) The area to be inspected is cleaned and visually inspected.
(c) Surface preparations include the diving preparations, but for a magnetic particle test the following checks should also be made:
  (i) The ink is made up to the required specification, or the ink to be used is tested to see that it meets the specifications.
  (ii) The ink dispensing unit is tested to see that the ink flows from the dispensing nozzle freely.
  (iii) The diver-inspector checks that the equipment supplied to produce the magnetic field is the type called for in the specification and that it is functioning correctly.

(iv) The diver-inspector also tests the strength of the u/v light to see that it has not deteriorated below the standard required for the test (if BS is used, then an illumination of 50 lux white light equivalent is called for).

(v) The diver-inspector ensures that the instructions for recording observed defects are clearly understood.

(d) When the equipment is positioned on the site to be tested, demagnetization is carried out if this is called for in the specification.

(e) The diver-inspector checks that the magnetic field being used is adequate by using a field indicator such as the Burmah–Castrol strip.

(f) The diver-inspector applies the ink and inspects using the u/v light.

(g) If a crack or other defect is detected, the diver-inspector brushes away the particles and retests to confirm the original observation.

(h) The diver-inspector reports and records the observations made during the test in the manner outlined in the specification.

(i) Demagnetization is carried out, as required.

10.11   THE SENSITIVITY OF MPI

The sensitivity of an MPI inspection is the effectiveness with which the test will discover defects and cracks in the material. This will depend in the main on two factors: the operator (or in this case, the diver-inspector), and the equipment and conditions.

For maximum efficiency of the operation, the diver-inspectors should, as well as being as comfortable and well equipped to dive as possible, have confidence in their ability to use the equipment for a particular test, in their own ability to detect defects with it, and in the value of their contribution to the efficiency and safety of the plant he is inspecting. The last three are obtained from the competence and confidence imparted to the inspector by good training and an active involvement in the inspection function.

The sensitivity of the magnetic particle test will depend on several factors, some of which will be within the control of the inspector and others not. The sensitivity of detection depends ultimately on the contrast that can be produced between the defect and its surroundings and the definition which tells us the size, shape and orientation of the defect.

Figure 10.24 shows some of the factors that will affect the performance and, as with the chain, it will only be as good as its weakest link.

So far as the contrast is concerned, we would outline the following points for consideration:

(a)   Surface condition

The effectiveness of the cleaning procedures to produce a bright finish is an

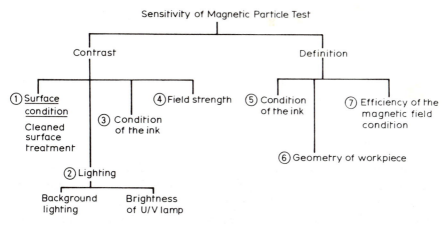

Fig. 10.24   Sensitivity of magnetic particle test.

important factor here. A paint background would help if coloured magnetic particles are being used.

(b)   Lighting

If the ambient lighting is too high, for example because of bright sun near the surface, the tests should be done at night. The diver-inspector should monitor regularly the intensity of the u/v lamp to detect any deterioration in brightness and so a decrease in the efficiency of detecting cracks.

(c)   Condition of the ink

Important factors here include the best colour to attract the attention of the inspector and the right size of particle. At the moment, this is tightly controlled, but perhaps a larger size particle would give better contrast to the diver-inspector. The physical condition of the ink is also important. It should be a finely divided suspension of particles that is delivered to the workpiece and so the diver-inspector should check to see that the agitator is working. This should be a regular part of the routine.

(d)   Field strength

This must be high enough to hold the ink to the surface defect.

The following factors are important when considering definition:

(a)   Condition of the ink

This appears twice in Fig. 10.24, as the definition might be improved if the defect outline were picked out in fine magnetic particles.

(b)   Geometry of the workpiece

In the main, MPI has to be carried out by diver-inspectors on welds that are parts of complicated nodes in very difficult positions in a hostile environment. This might be the most significant weakness in the system and methods of improving the procedures, back-up for divers at the work position and redesign of the equipment with this in mind will bring about a marked improvement in defect detection.

(c)   Efficiency of the magnetic field conditions

The effectiveness with which the inspector can set up the magnetic field conditions is the heart of the magnetic inspection. Factors such as current flow, field direction, electrical contact of the prods, coil fill factor, etc. need to be considered carefully.

10.12   MPI OF AN OFFSHORE STRUCTURE BY A MAJOR NORTH SEA CONTRACTOR, SUMMER 1984

10.12.1   Background

The inspection was undertaken on a selected node joint at the −120 m (394 ft) level. At this depth, there is no sunlight penetration, so low ambient light was assured. The equipment used was the OIS MPI unit, which utilizes coils. The diver-inspector was qualified to the required standard and reported directly to the inspection engineer in the control room. All equipment was tested and calibrated prior to the inspection.

10.12.2   Inspection method

Prior to inspection, the joint was cleaned by water jet incorporating grit injection, to a matt, clean metal finish. This extended 100 mm (3.9 in) either side of the weld. A magnetic tape was placed round the joint, clear of the HAZ (heat affected zone), for identification purposes. The subsea unit was then lowered into position down the downline. The diver-inspector positioned the coils around the member and then applied first the flux and then the ink.

During the inspection phase, the diver-inspector made a verbal report to the inspection engineer on the surface, in the form of a running commentary. The entire inspection was monitored by a remotely operated vehicle (ROV) and the surface personnel viewed the live video during the inspection.

On completion of the inspection, the equipment was derigged, and the inspection engineer recorded the data directly onto a computer record. Subsequently, the computer proformas were collated and filed as a hard copy back-up to the computer record.

11 Ultrasonic flaw detection and thickness measurement

11.1 ULTRASONIC RANGE

Ultrasonic testing depends on the way in which sound waves pass through the material under test. Ultrasonic sound waves – that is, high-frequency sound waves – are used because it is at these frequencies that the sound wave travels furthest with the minimum loss of energy in solids. This contrasts with the conditions in liquids, where the lower sonic frequencies penetrate with the minimum loss of energy. The loss of energy of a signal (any signal) is known as the attenuation of the signal. The ultrasonic sound wave range of frequency is from 20 kHz to 10 MHz. For ultrasonic non-destructive testing, the frequencies between 0.5 MHz and 10 MHz are used. As ultrasonic sound cannot be seen or heard or sensed in any other way by human beings, very high levels of signal energy can be used, which would be unbearable if they were used in the audible range.

11.2 FREQUENCY OF THE WAVE

The signal used rises and falls and reverses direction. This pattern is repeated over and over again and the number of times this is repeated in one second is the frequency of the signal. The basic unit of frequency is the Hertz, abbreviated to Hz, and one Hertz is one complete cycle of an event in one second. For example, with a varying voltage of frequency 1 Hz, the voltage would increase from zero to maximum positive value, decrease to zero, increase to maximum negative value and finally decrease to zero, all in 1 second (see Fig. 11.1).

$$\text{Frequency } (f) = \frac{\text{Number of cycles}}{\text{Time for that number of cycles}}.$$

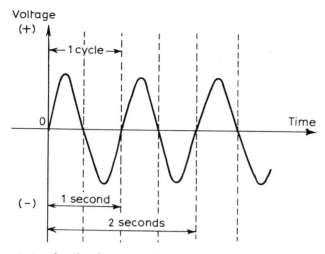

Fig. 11.1   A signal with a frequency of 1 Hz.

The time for one cycle is known as the periodic time (*P*) and is measured in seconds. Therefore,

$$f = \frac{1}{P} \text{ Hertz}$$

To calculate the frequency of the signal shown in Fig. 11.2:

$$\text{Frequency } (f) = \frac{\text{Number of cycles}}{\text{Time for that number of cycles}}$$

Number of cycles $= 2.5$

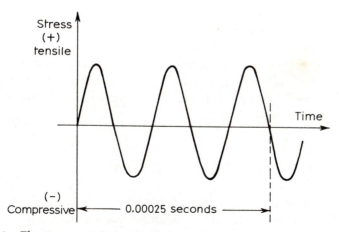

Fig. 11.2   The stress variation with time at a point in a solid subjected to ultrasonic impulse from a transducer.

Time taken for these cycles $= 0.00025$ second

Therefore, the frequency (f)

$$= \frac{2.5}{0.00025} \text{ Hertz}$$

$$= 10\ 000 \text{ Hertz.}$$

To calculate the periodic time (this is the time for one cycle):

$$P = \frac{0.00025}{2.5} \text{ second} = 0.0001 \text{ second.}$$

It is not usual to use all the noughts with the numbers, and the following is the way in which we reduce the writing:

Number	Prefix	Symbol
1 000 000 $= 10^6$	mega	M
1 000 $= 10^3$	kilo	k
1 $=$ 1		
0.001 $= 10^{-3}$	milli	m
0.000001 $= 10^{-6}$	micro	μ

The frequency calculated above would normally be written as 10 kHz and the period of the signal as 100 μsec.

11.3   SPEED OF THE WAVE

The discussion so far has been about what one point in the material experiences as the ultrasonic wave passes by, but of course the wave itself is passing along through the material. (As with a surface wave on the water, the water at any point goes up and down, but as well as this, the wave travels forwards.) Ultrasonic waves travel through a solid at the speed of sound for any given type of wave in a given material.

11.4   TYPES OF ULTRASONIC WAVE

The sound wave propagates through the material (liquid, solid or gas) by causing the atoms to oscillate as the wave front passes through it. There are two types of wave that propagate through the solid material and three types of surface wave.

11.4.1   Waves that propagate through solids

These include the following.

(a)   Longitudinal or compression waves

The symbol used to denote this type of wave is L. Thus, V_L is the velocity of the propagation of these longitudinal or compression waves. With this type of wave propagation, the direction of the oscillation of the atoms is the same as the direction of the wave propagation (see Fig. 11.3, which shows the wave at a given instant in time; as with all other types of wave, it is moving to the right as time progresses).

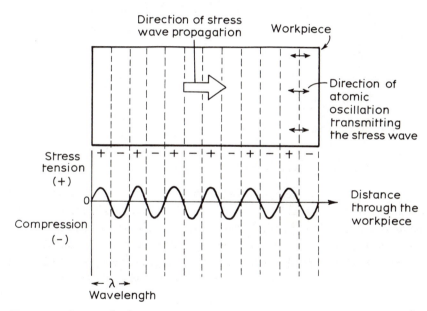

Fig. 11.3   Longitudinal or compression waves.

(b)   Shear or transverse waves

The symbol used to denote this type of wave is T, so that V_T is the velocity of propagation of these transverse or shear waves. With this type of wave propagation, the direction of oscillation of the atoms is at right angles to the direction of motion of the propagating wave (see Fig. 11.4).

As with Fig. 11.3, the shear wave is shown at a given instant in time. The main difference is in the movement of the atoms: whereas in the compression waves the atoms are pulled apart and compressed in the direction that the wave is moving, in the shear waves the atoms are pushed past each other — that is, sheared — see Fig. 11.5.

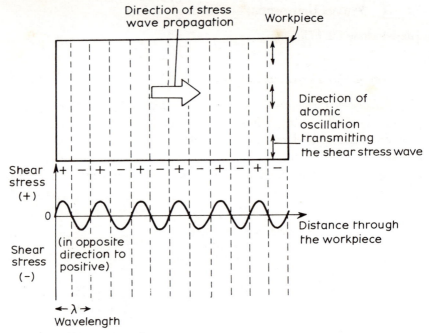

Fig. 11.4   Transverse or shear waves.

11.4.2   Surface waves

The main type of wave here is the Rayleigh wave. These waves travel only on the surface of the material. The atomic motion of the wave is elliptical, with the major axis of the ellipse perpendicular to the surface, thus resembling a surface wave on the water. The wave can be easily damped out if the surface is in contact with either a solid or a fluid. On land, this type of wave cannot be used in an immersion technique, so it is unlikely to be used under the sea.

(a)   Lamb waves

These are generated when the thickness of the material is comparable to the wavelength of the Lamb wave. There are two main types of Lamb wave, symmetrical and asymmetrical, and each of these has a series of modes. In this respect, they differ from those already mentioned, which propagate only in one mode. On land, Lamb waves are applied for testing thin wall tubing and for laminar defects which lie very close to the surface of the part. They are also used to test the quality of bond in laminate materials.

(b)   Love waves

These are waves that travel on the surface of the material without a vertical component of displacement, analogous to a surface compression wave motion.

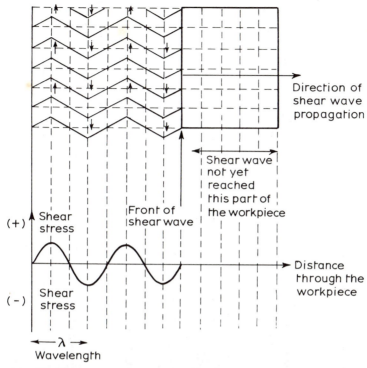

Fig. 11.5   Diagrammatic representation of the movement of the atoms as a shear wave passes through the material. Dotted lines represent the position of the rows of atoms before being subjected to the shear wave.

11.5   VELOCITY OF ULTRASONIC WAVES

The interpretation of ultrasonic data is dependent on a knowledge of the velocity of the wave propagation. For measurement purposes, it is essential that the velocity of the sound wave is the same for different samples of the same material. A summary of the wave velocity of the various waves just discussed is given in Fig. 11.6.

Methods for calculating some of the single value velocities are given in Fig. 11.7. Symbols used:

$$V = \text{velocity of sound (m s}^{-1})$$
$$E = \text{elastic modulus (N m}^{-2})$$
$$G = \text{shear modulus (N m}^{-2})$$
$$\rho = \text{density (kg m}^{-3})$$
$$\mu = \text{Poisson's ratio}$$

In some books, C is used as the velocity of sound. A table of the related properties for common materials is given in Fig. 11.8.

Name of wave motion	Symbol for velocity	Conditions	Atomic motion	Is the velocity of single value?	Comments on subsea use
Longitudinal or compression	V_L	Passage through large bulk of material	Longitudinal – that is, in the direction of wave propagation	Yes	Used for thickness and lamination measurement
Shear or transverse	V_T	Passage through large bulk of material	Transverse – that is, at right angles to the direction of wave propagation	Yes	Will be used for defect sizing
Rayleigh, often referred to as surface waves	V_S	Semi-infinite free surface	Compound – that is, motion up and down as well as in the direction of the waves	Yes	Not used under water
Lamb or 'plate' waves	V_m	Thin sheet	Compound	No	Not used under water
Thin rods	V_o	Small diameter bars	Compound	Yes	Not used under water

Fig. 11.6 Summary of wave velocity.

Type of wave	Velocity of sound
Compression or longitudinal	$V_{\mathrm{L}} = \sqrt{\left[\dfrac{E}{\rho} \dfrac{1-\mu}{(1+\mu)(1-2\mu)} \right]}$
Shear or transverse	$V_{\mathrm{T}} = \sqrt{\left[\dfrac{E}{\rho} \dfrac{1}{2(1+\mu)} \right]}$
	$V_{\mathrm{T}} = \sqrt{(G/\rho)}$

Fig. 11.7 Formulae for calculating acoustic velocities.

11.6 ULTRASONIC WAVELENGTH (λ)

The wavelength tells us how far the ultrasonic stress wave moves forward for one complete stress cycle, and needs to be considered when determining the maximum sensitivity of the set-up. It is calculated using the following equation:

$$\text{Wavelength } \lambda = \frac{\text{Velocity of sound}}{\text{Frequency of the ultrasonic wave}}$$

For example, to calculate the wavelength for a 20 kHz ultrasonic compression wave signal in aluminium:

$$\text{Wavelength } \lambda = \frac{6350 \ (\mathrm{m \ s^{-1}})^*}{20 \ 000 \ \text{cycles s}^{-1}}$$

$$\lambda = 0.3175 \text{ metres.}$$

*from Fig. 11.8.

Now calculate the wavelength for a 5 MHz compression wave in steel (1.17×10^{-3} m or 1.17 mm).

What would the wavelength be if it were a shear wave? (6.46×10^{-4} m or 0.646 mm).

Because the wavelength depends on the velocity of sound, the wavelength of the ultrasonic signal will change when it passes from one material into another. For example, when an ultrasonic compression wave passes from Perspex into steel (Perspex would be the material used for the shoe of the transducer and steel the workpiece), the wavelength in steel would be larger than that in the Perspex as the velocity of sound in Perspex is smaller than the velocity in steel (see Fig. 11.9).

Material	Acoustic velocity (m s⁻¹)				Density (kg m⁻³)	Modulus (GN M⁻²)	Modulus (GN M⁻²)
	V_L	V_T	V_S	V_0	ρ	E	G
Aluminium	6 350	3 100	2 900	5 100	2 710	70.8	26.5
Araldite	2 500				1 200		
Concrete	4 600				2 000		
Brass (naval)	4 430	2 120	1 950	3 490	8 100	98.4	36.4
Copper	4 660	2 260	1 930	3 710	8 900	122.7	45.5
Lead (pure)	2 160	700	630	1 200	11 400	16.1	5.6
Air	333				1		
Bronze, phosphor (5%)	3 530	2 230	2 010	3 430	8 860		
Lead, antimony (6%)	2 160	810	740	1 370	10 900		
Steel (structural)	5 940	3 250			7 850	213.3	82.9
Steel	5 850	3 230	2 790	5 170	7 800		
Steel – stainless 302	5 660	3 120	3 120	4 900	8 030		
Steel – stainless 410	7 390	2 990	2 160	5 030	7 670		
Water	1 490				1 000		
Motor oil	1 740				870		
Transformer oil	1 380				920		
Nylon	2 620	1 080			1 100	3.59	1.28
Polyethylene	2 340	925			940	2.26	0.80
Perspex (Plexiglass)	2 730	1 430			1 180	6.33	2.41

Fig. 11.8 Table of properties.

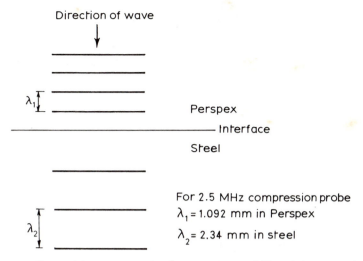

Fig. 11.9   Compression wave passing from one material (Perspex) into another (steel).

11.7   FURTHER INFORMATION ON ULTRASONIC PROPERTIES IN MATERIALS

11.7.1   Acoustic impedance (z)

This is a useful parameter of the material if the pressure or stress magnitude of the ultrasonic wave is to be determined. The acoustic impedance (z) of a material is the product of density (ρ) and acoustic velocity (V). That is,

$$z = \rho \times V$$

(For a compression wave, V would be V_L.)

The pressure or stress magnitude (p) of the ultrasonic wave is proportional to the acoustic impedance (z), the frequency (f), and the displacement of amplitude of the atoms (d). That is,

$$p = 2\pi z f d.$$

11.7.2   Acoustic attenuation

This is the reduction of the energy of the ultrasonic wave as it passes through the material. Large amounts of attenuation will reduce the 'penetrability' of the ultrasonic wave and a loss of the back wall reflection could be caused by this effect. The main physical phenomena that attenuate the ultrasonic signal are scattering, diffusion, viscous damping losses and relaxation losses. The first two are due to the effect of stressing the atoms in the material.

11.7.3   The decibel system

It is usual to measure the performance of an amplifier or attenuator in terms of the ratio of one signal to another; e.g. input to output; in the case of the amplifier, the output is larger than the input, and for the attenuator, the reverse is the case. It is therefore consistent to measure the attenuation of a signal using this technique. It is not an absolute measurement, but gives the relationship for the decay of the energy in the signal. Consider the decay of the power in a signal as it passes through a piece of material (see Fig. 11.10).

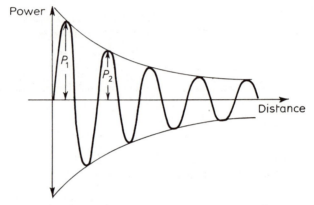

Fig. 11.10   Power in the ultrasonic signal with distance into the material.

The amount the signal attenuates is given by the curve that joins up the peaks of the graphs and the attenuation in decibels (dB) is given by:

$$10 \log \frac{P_1}{P_2} \text{ dB}$$

P_1 and P_2 can be any successive two peaks.

The ultrasonic signal from the probe is, however, a voltage and the ratio of the powers

$$\frac{P_1}{P_2}$$

is given by the ratio of the square of the voltages:

$$\frac{V_1^2}{V_2^2}$$

because power $P = I$ (amps) $\times V$ (volts) and I (amps) = volts/resistance. Therefore, $P \alpha (V)^2$.

This, then, gives the ultrasonic attenuation as:

$$10 \log \frac{(V_1)^2}{(V_2)^2} \text{ dB}$$

which reduces to:

$$20 \log \left(\frac{V_1}{V_2}\right) \text{ dB}.$$

Therefore, in ultrasonics, we calculate the attenuation in dB from the expression:

$$20 \log \left(\frac{V_1}{V_2}\right)$$

For example, what is the attenuation of the ultrasonic signal between the first and second echo for the signal shown in Fig. 11.11?

$$\text{Attenuation (dB)} = 20 \log \left(\frac{12}{8}\right)$$

$$= 20 \log (1.5)$$

$$= 20 \times 0.1761$$

$$= 3.52 \text{ dB}.$$

What would the attenuation be if V_2 were exactly half V_1 and what would it be if V_2 were one tenth V_1? (6 dB, 20 dB.)

The last two results calculated are most important and the reader should

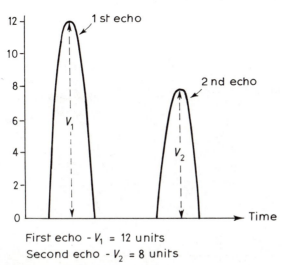

First echo - V_1 = 12 units
Second echo - V_2 = 8 units

Fig. 11.11   Attenuation of ultrasonic signal between two echoes.

remember that if the signal is halved that is a 6 dB drop, and if the signal is one tenth then that is a 20 dB drop.

11.7.4   Ultrasonic wave directions

So far we have seen that the ultrasonic wave travels at a known speed in straight lines. In order to predict the direction in which the wave travels once it is inside the material, we need to know what happens when the wave meets an interface. An interface is the boundary between two materials of differing properties (e.g. water and steel, Perspex and steel, water and air, etc.), and will include the outside edges of a component, often referred to as the back wall, or the surface of a crack or edge of a porosity bubble. At these interfaces, the direction after meeting the interface will be determined by the law of reflection for the wave moving in the initial material and the law of refraction (Snell's law) for the wave that passes into the second material. These laws are the same as those for light wave motion.

(a)   Law of reflection

This states that the angle the reflected wave makes with the normal to the interface from which the wave is being reflected is the same as the angle that the incident wave makes with the same normal (see Fig. 11.12). When the angle of incidence α is 0 then the reflected wave angle is also 0, so the wave is reflected back along the incident direction and as the wave is travelling in the same material, there will be no change in wavelength of the signal or the type of wave. This is the ideal condition for thickness measuring meters using ultrasonic compression waves.

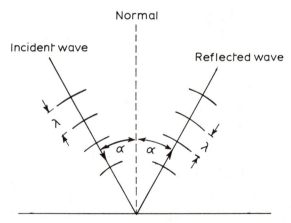

Fig. 11.12   Reflection of ultrasonic wave from an interface.

(b) Law of refraction

At an interface, part of the ultrasonic wave is reflected and the rest will pass on into the second material. The path in this material will still be a straight line, but the direction of this wave will not be continuous with the direction of the incident wave, as it will have been turned through an angle which can be determined by Snell's law (see Fig. 11.13).

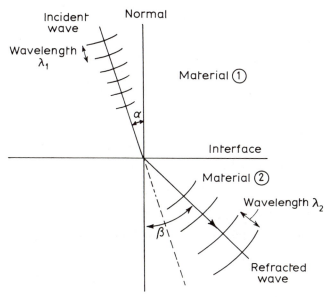

Fig. 11.13   Snell's law.

Snell's refraction law states that:

$$\frac{\sin \alpha}{\sin \beta} = \text{constant}$$

α need not necessarily be larger than β as shown in Fig. 11.13. The value of the constant is determined by the properties of material 1 and material 2, and the wavelength will have changed in going from material 1 to material 2.

The constant in Snell's law is the ratio of the velocity of sound in the two materials, so that we now write Snell's law as:

$$\frac{\sin \alpha}{\sin \beta} = \frac{\text{Velocity of sound in 1}}{\text{Velocity of sound in 2}}$$

As there are two velocities of sound in a material, V_L and V_T, if the incident wave can cause these then there will be two resulting refracted waves. An incident compression (longitudinal) wave over a range of angles can produce two stress waves at the interface in material 2, as it has a component parallel to the interface given by sin α which will produce a shear wave in material 2 (see Fig. 11.14). Both these resulting waves in material 2 will be turned from the incident direction by an angle determined by Snell's law and the velocity of sound in the two materials (see Fig. 11.15).

The angle that the compression wave makes with the normal in material 2 is determined from Snell's law, which gives:

$$\frac{\sin \alpha}{\sin \beta_L} = \frac{V_{L1}}{V_{L2}}$$

Where V_{L1} is the velocity of sound in 1 and V_{L2} is the velocity of sound in 2. Therefore, β_L is the angle whose sine is equal to

$$\frac{V_{L2}}{V_{L1}} \sin \alpha$$

The angle that the resulting shear wave makes with the normal in material 2 is again determined using Snell's law, except that this time we use the velocity of sound of the shear wave (V_{T2}) in the equation, so that:

$$\frac{\sin \alpha}{\sin \beta} = \frac{V_{L1}}{V_{L2}}$$

so that the angle that the resulting shear waves makes with the normal in material 2, β_T is the angle whose sine is

$$\frac{V_{T2}}{V_{L1}} \sin \alpha$$

Example: Determine the resultant wave angles and wavelengths when an

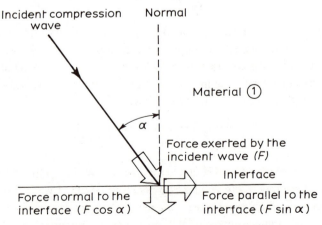

Fig. 11.14   Force exerted by the incident wave on the interface.

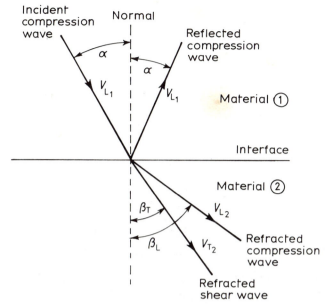

Fig. 11.15   Reflected and refracted shear wave and compression wave at an interface of two materials.

ultrasonic compression wave of 50 kHz travelling in Perspex meets an interface with steel at an angle of 10° to the normal.

From Fig. 11.8:

$$V_L \text{ in Perspex is } 2730 \text{ m s}^{-1}$$

$$V_L \text{ in steel is } 5850 \text{ m s}^{-1}$$

$$V_T \text{ in steel is } 3230 \text{ m s}^{-1}.$$

The incident compression wave in Perspex will, on meeting the interface, produce three ultrasonic waves (see Fig. 11.16):

 (i) A reflected compression wave in the Perspex
 (ii) A refracted compression wave in the steel
(iii) A refracted shear wave in the steel.

(c)   Reflected compression wave in the Perspex

The law of reflection tells us that the angle of reflection equals the angle of incidence, so that the reflected wave makes an angle of 10° to the normal.

The wavelength of the reflected wave will be the same as the incident wave, as they are both compressive and travelling at the same speed. This is calculated from the following equation:

$$\lambda = \frac{V}{f}$$

$$\lambda = \frac{2730 \text{ m s}^{-1}}{50\ 000 \text{ Hz}} = 0.0546 \text{ m or } 54.6 \text{ mm}$$

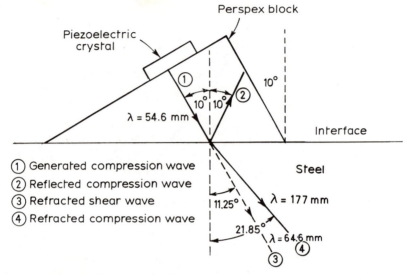

Fig. 11.16   Diagram of wave configuration from calculation.

(d)   Refracted compression wave in the steel

Snell's law gives:

$$\frac{\sin \alpha}{\sin \beta_L} = \frac{V_{L1}}{V_{L2}}$$

Therefore,

$$\sin \beta_L = \frac{V_{L2}}{V_{L1}} \sin \alpha$$

$$\sin \beta_L = \frac{5850}{2730} \sin 10°$$

$$\sin \beta_L = \frac{5850}{2730} \, 0.1736$$

$$\sin \beta_L = 0.3721$$

$$\beta_L = 21.85°.$$

The wavelength of the compression wave in steel:

$$\lambda = \frac{5850 \text{ m s}^{-1}}{50\ 000 \text{ Hz}}$$

$\lambda = 0.117$ m or 117 mm.

(e) Refracted shear wave in the steel

Snell's law also gives:

$$\frac{\sin \alpha}{\sin \beta_T} = \frac{V_{L1}}{V_{T2}}$$

$$\sin \beta_T = \frac{V_{T2}}{V_{L1}} \sin \alpha$$

$$\sin \beta_T = \frac{3230}{2730} \sin 10°$$

$$\sin \beta_T = \frac{3230}{2730} 0.1736$$

$$\sin \beta_T = 0.2054$$

$$\beta_T = 11.85°.$$

The wavelength of the shear wave in steel:

$$\lambda = \frac{3230 \text{ m s}^{-1}}{50\ 000 \text{ Hz}}$$

$\lambda = 0.0646$ m or 64.6 mm.

It will be seen that by following through the last calculation there will be two critical angles. These are angles of inclination at which the ultrasonic wave will not pass into material 2 from material 1.

The first critical angle is when the refracted compression wave angle is 90°. This, of course, means that the only wave transmitted is the shear wave.

As the incident angle to the normal increases, the second critical angle is reached. At this condition, the shear wave is passed along the interface as a surface wave.

Example: Show that for the data given in the previous example, the first critical angle is 27.8° and the second critical angle is 57.7°.

To show that the first critical angle is 27.8° for steel: from Fig. 11.8:

$$V_L \text{ Perspex} = 2730 \text{ m s}^{-1}$$

$$V_L \text{ steel} = 5850 \text{ m s}^{-1}$$

Snell's law:

$$\frac{\sin \alpha}{\sin \beta} = \frac{V_{L1}}{V_{L2}}$$

$$\sin \alpha = \frac{V_{L1} \times \sin \beta}{V_{L2}}$$

$$\sin \alpha = \frac{2730 \times 1}{5850}$$

$$\sin \alpha = 0.4664.$$

From tables, $27.8° = 0.4664$. Therefore, the first critical angle for steel must be $27.8°$.

To show that the second critical angle is $57.7°$ for steel: from Fig. 11.8:

$$V_L \text{ Perspex} = 2730 \text{ m s}^{-1}$$

$$V_T \text{ steel} = 3230 \text{ m s}^{-1}$$

Snell's law:

$$\frac{\sin \alpha}{\sin \beta} = \frac{V_{L1}}{V_{T2}}$$

$$\sin \alpha = \frac{V_{L1} \times \sin \beta}{V_{T2}}$$

$$\sin \alpha = \frac{2730 \times 1}{3230}$$

$$\sin \alpha = 0.8452$$

From tables, $\sin 57.7° = 0.8452$. Therefore, the second critical angle for steel must be $57.7°$.

11.7.5   Test frequency

The test frequency used for flaw detection in land-based equipment varies with different applications. The following are some examples of frequency and application.

Applications	Frequency range
Concrete, wood, natural rocks	25–100 kHz
Coarse grain metal structures (e.g. cast iron, copper, stainless steel)	200 kHz–1 MHz
Finer grain metal structures	400 kHz–5 MHz
Plastics	200 kHz–2.25 MHz
Forgings	1–10 MHz
Welds	1–2.25 MHz

11.8   ULTRASONIC TRANSDUCERS

The word transducer has a very general meaning which is any device which transforms energy from one form to another.

In the case of the ultrasonic transducer, it transforms high frequency electrical signals to the same high frequency mechanical signals and vice versa. There are two possible types of device for doing this. At the low frequency end of the scale (below 100 kHz) a magnetostrictive device is used, and above this range piezoelectric devices are used. At the moment, ultrasonic non-destructive testing uses mostly piezoelectric devices, while magnetostrictive devices are widely used in sonar and underwater signalling.

Magnetostrictive devices use the expansion and contraction of magnetic materials under the influence of a varying magnetic field, to generate a mechanical pulse from a magnetic signal. The device also works in reverse, the input of a mechanical signal generating a magnetic signal.

The piezoelectric transducer makes use of the property that certain crystalline and ceramic materials have, whereby, if the material is subjected to a mechanical extension or contraction, it generates an electrical signal related to the size of the mechanical input. This effect is also reversible, although the efficiency of the conversion is dependent on the type of use. Early transducers were made from quartz and Rochelle salt crystals, but now they are made from a range of synthetic crystalline and ceramic materials that include barium titanate, lithium sulphate and potassium zirconium titanate.

The transducer's crystal is fitted with electrodes (see Fig. 11.17) to apply the voltage across the crystal if it is acting as a transmitter, or to take the voltage signal from it if it is acting as a receiver. The crystal is attached to the case by the mounting, which acts not only as a fixing but also as a backing to the crystal.

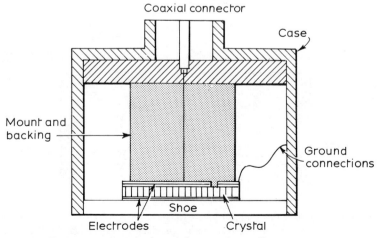

Fig. 11.17   Basic arrangement of a single crystal transducer.

The backing member has a significant influence on the transducer's performance. Its impedance controls the bandwidth of the transducer and the damping of the crystal's ring. For maximum damping, the impedance of the backing member must equal the impedance of the crystal. Also, it must have the ability to attenuate completely the sonic energy transmitted from the back of the piezoelectric element so that this signal is prevented from returning to the back of the piezoelectric crystal, thereby producing an unwanted signal.

The shoe is added to protect the crystal from physical damage or wear, and also from the environment. Most transducers are completely sealed units. The shoe can be shaped to act as a lens. In this respect, sound waves behave like light waves and acoustic lenses are designed in a similar manner to light lenses.

The shoe should absorb as little of the sound as possible and so its impedance is usually selected to be between the crystal and the material with which the transducer is in contact. Shoes are often constructed of Perspex.

Also, a probe that operates in a shear mode can be produced from a piezoelectric crystal that operates in a longitudinal mode, by having a wedge-shaped shoe (see Fig. 11.18). The wedge shape directs the longitudinal wave at an angle to the surface of the test part. Depending on the angle of incidence, this will produce partial or total conversion of the longitudinal wave into a shear wave, as seen previously.

11.8.1   Types of transducer

(a)   A normal probe

This transmits longitudinal or compression waves, and separate transmitters

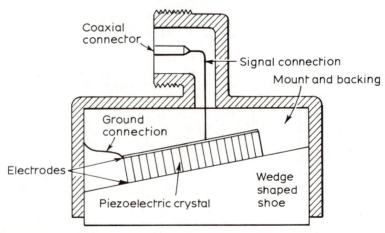

Fig. 11.18   Angled probe using a piezoelectric crystal in the longitudinal or compression mode, the wedge producing a shear wave in the material.

and receivers (each with one crystal) are made with this type of configuration (see Fig. 11.17).

(b) Single crystal probes

These are probes where a single piezoelectric crystal transmits and receives the ultrasonic signal. The acoustic characteristics of this transducer are very different from the single function transducer. The crystal has to transmit the signal, stop ringing and await the returning echo, so the natural frequency of the crystal needs to be very much different from that of the ultrasonic frequency being used for the test.

(c) Twin probes (TR probes)

This is a type of transducer where the transmitter and receiver are mounted in the same housing, although electrically and acoustically they are separated (see Fig. 11.19). The acoustic insulation is generally a thin layer of cork. This is the type of transducer used in the Baugh and Weedon Sea Probe and the Krautkramer D meter. It is also in general use as a probe for many types of flaw detector.

(d) Angle beam probes

This type of probe (see Fig. 11.18) produces an ultrasonic beam which is introduced to the material at an angle to the interface, and not perpendicularly, as in the case of the normal beam probe (see Fig. 11.17). As can be seen from the section on Snell's law (Section 11.7.4(b)), this type of probe can produce shear

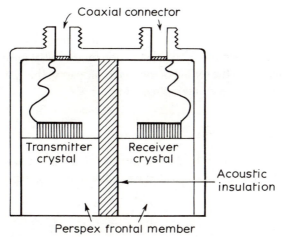

Fig. 11.19   Twin or TR probe.

and compression waves, just shear waves, or surface waves, depending on the angle of the probe.

11.9   COUPLANTS

In air, a contact test cannot be carried out without the use of a suitable coupling agent between the transducer and the material. This is because the small mechanical pulses cannot travel across the small air gap that exists between the two surfaces, because of the mismatch in acoustic impedance between the shoe of the transducer and the air. For land-based tests, liquids, greases or pastes are therefore used. Under the sea, we have the equivalent of the land-based immersion technique, and transmission across the gap between the workpiece and the transducer is no problem. The surface of the workpiece must, of course, be free from marine fouling as this would attenuate and scatter the ultrasonic signal so that it would not even enter the workpiece.

11.10   THE ULTRASONIC BEAM

In order to use the passage of the ultrasonic wave through the material to search for defects, we need to be able to visualize the effective volume of material covered, and as with light from a torch, we consider the concept of an ultrasonic beam.

The shape of the beam can be considered as a short cylindrical portion containing the dead zone and near zone followed by a conically shaped far zone (see Fig. 11.20).

11.10.1   The dead zone

This is the region just ahead of the probe. Reflections from defects in this region cannot be observed because of the obstruction by the initial pulse. This will be very much affected by the setting of the electronic part of the system; e.g. pulse length, amplifier recovery time, gain setting.

11.10.2   The near zone

This is the name given to the parallel portion of the beam just ahead of the transducer and its length is calculated using the following expression:

$$\frac{D^2}{4 \times \lambda}$$

where D is the crystal diameter and λ is the wavelength.

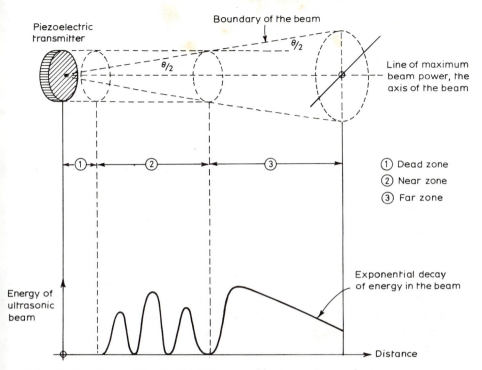

Fig. 11.20   Shape of the ultrasonic beam and beam energy envelope.

This equation calculates the length of the near zone from the crystal face in one material. If, however, the near zone is not contained within the shoe, then an allowance must be made for the change in near zone length as the ultrasonic beam travels from one material into another.

11.10.3   The far zone

In the far zone, the boundary of the cone is defined by attenuation of the maximum power at any section to a given level, either 6 dB or 20 dB.

The cone angle defined as $\theta/2$ in Fig. 11.20 is given by the following equation:

$$\sin(\theta/2) = \frac{k\lambda}{D}$$

where again λ is the wavelength, D is the diameter of the crystal and k is a constant factor. For 6 dB, drop $k=0.56$ and for 20 dB, drop $k=1.06$. In this region, the energy of the ultrasonic beam decays exponentially. It is generally this part of the beam that we use for thickness measurement, flaw detection and measurement.

11.11   PRINCIPLES OF ULTRASONIC TESTING

There are two basic principles of ultrasonic testing. The first is based on the detection of a decrease in energy of the ultrasonic beam due to absorption by the flaw. This often involves the transformation of energy due to the internal friction of the defect or a failure to transmit energy across the air gap of the defect (see Fig. 11.21). This is sometimes referred to as the shadow method.

The second principle of ultrasonic testing is based on the reflection of energy from a flaw or interface (see Fig. 11.22).

11.12   ULTRASONIC TEST SYSTEMS

From the above paragraphs, it will be seen that an ultrasonic test system should be able to measure either the amplitude of the signal if the first type of test is used or the time required for the ultrasonic signal to travel between specific interfaces. A versatile test system will measure both the parameters at the same time, although for thickness measurement the main use of ultrasonics subsea at the moment is for the measurement of the time the ultrasonic signal takes to travel between specific interfaces.

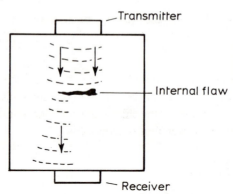

Fig. 11.21   Flaw detection by decrease in energy of the ultrasonic wave.

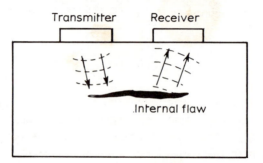

Fig. 11.22   Flaw detection by reflection of the ultrasonic wave.

A test system for ultrasonic thickness measurement is based on a pulse flaw detector circuit, a simple arrangement for which is shown in Fig. 11.23. The technique employed is similar to that used in some systems of radar and depth-sounding. A short duration electrical pulse is generated and fed to the transmitting transducer. This rings mechanically at an ultrasonic frequency due to the piezoelectric cyrstal's conversion of electrical energy to mechanical energy. While the equipment is in use, the transducer produces short bursts of ultrasonic waves which pass through the material. This signal is then picked up by the receiving transducer crystal, which converts the mechanical ultrasonic pressure wave back into an electrical signal. The timing circuits then measure the interval between the transmitted and received signals. This cycle of events is repeated often enough for an essentially continuous indication of the information to be obtained, but the time lapse between cycles is sufficient to allow the ultrasonic waves to die away.

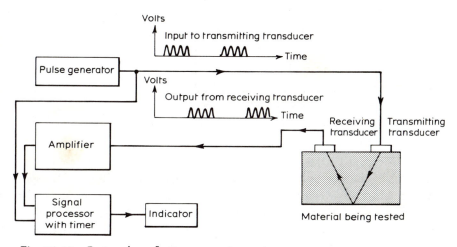

Fig. 11.23   Basic pulse reflection measuring system.

Several possible types of unit can be used in the system, depending on the method employed for timing and indicating. For ultrasonic flaw detection, the indicator is generally a cathode ray oscilloscope as this presents all the information available in the echo signal. For thickness measurement, it is generally more convenient to present the information on a digital meter, which often presents the data directly as the thickness of the material and not as time.

11.13   ULTRASONIC THICKNESS MEASUREMENT

The thickness of a material can be measured using ultrasonic wave propagation because certain types of these waves travel at a constant speed through a material, because they travel in straight lines, and because a portion of the wave is reflected when it meets an interface. Normally, longitudinal waves are used for thickness measurement.

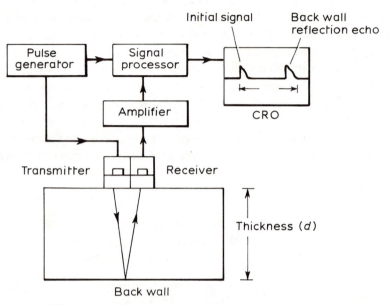

Fig. 11.24   Ultrasonic thickness measuring system.

Let us consider a thickness measuring system using a TR (twin) probe (see Fig. 11.24). The indicator unit is a cathode ray oscilloscope (CRO), and the spot moves across the screen at a fixed speed, so that distance measured across the screen represents time. This speed can be changed by the signal processor, which may be part of the CRO controls. The transmitted signal is indicated by the first rise on the screen of the CRO, and the second is the return signal, which is the reflection from the back wall. The distance between these two rises is the time (t) taken for the ultrasonic wave to travel across the material and return. The thickness of the material can then be calculated in the following way.

Figure 11.8 gives the speed of the longitudinal wave (V_L) in the material under test. For example, if the material is structural steel, the speed of the longitudinal wave (V_L) is 5940 m per second. The time taken for the wave to cross the material and back is t seconds, measured from the trace on the CRO screen. The distance travelled by the ultrasonic wave in this test is then given by the product of V_L and t. This distance is twice the thickness of the material as the wave has to travel there and back, so the thickness of the material (d) is given by the following expression:

$$d = 1/2(V_L \times t).$$

If (see Fig. 11.25) a through-transmission method is used, that is, where a

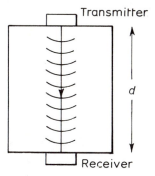

Fig. 11.25   Through-transmission arrangement.

receiver is placed on the opposite side from the transmitter, then the thickness
of the material is given by:

$$d = V_L \times t.$$

With specially designed underwater measuring equipment like the Baugh
and Weedon Sea Probe and the Krautkramer DMU (digital meter underwater),
the indicator of the system, a digital meter, gives a direct reading of thickness.
These instruments contain a rechargeable power supply. In the case of the Sea
Probe, power recharging is achieved by inductance through the case of the
instrument, thus avoiding the need for a penetration through the case. The
instrument also contains an electronic multiplier to convert the time
measurement into a thickness reading. Figure 11.26 shows a diagrammatic
representation of a typical underwater ultrasonic thickness measuring device.

In the Sea Probe, the value of V has been set in the instrument for steel and as
it is an instrument which is specially designed for underwater inspection of steel
structures, this is of positive advantage. If used on other materials, the
conversion of the reading for the new material is simply obtained by dividing
the reading of the meter by the value of V_L for steel and multiplying it by the V_L
for the new material. For example, if a measurement is made on an aluminium
structure, the meter reading has to be multiplied by:

$$\frac{6350}{5940}$$

in order to give the correct thickness of aluminium (value obtained from
Fig. 11.8).

In the case of the Krautkramer DMU there is provision to adjust this value
electrically to the correct reading of thickness for the material being inspected.
See Figs 11.27 and 11.28.

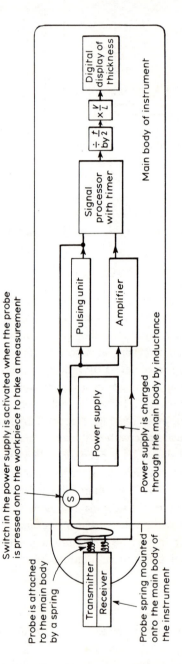

Fig. 11.26 A typical underwater thickness measuring device.

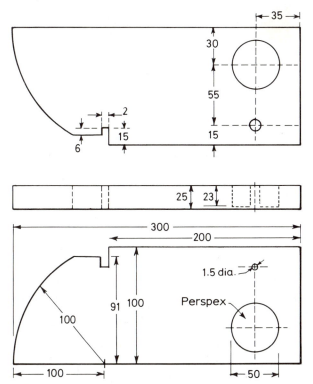

Fig. 11.27   IIW (International Institute of Welding) V1 calibration block (measurements in millimetres).

11.13.1   'A' Scan thickness measurement

Let us return in more detail to the cathode ray tube display of ultrasonic test information. With the 'A' Scan display, the baseline is time, which is calibrated to represent distance, and the vertical axis gives the magnitude of the returning signal. For thickness measurement, this signal will be the back wall echo.

Before using the equipment, a typical sequence of operations would be as follows:

(a) Check the power supply for portable sets, which are the majority of underwater sets, and see that the power supply is fully charged.
(b) Switch the equipment on and allow it to warm up, as this gives sufficient time for the electronics to stabilize.
(c) Adjust the focus and brilliance of the spot on the screen. If the spot is travelling quickly it will appear as a line on the screen.
(d) Using the delay control, find the initial pulse.
(e) Position the course range control to 100 mm, or an appropriate range setting.

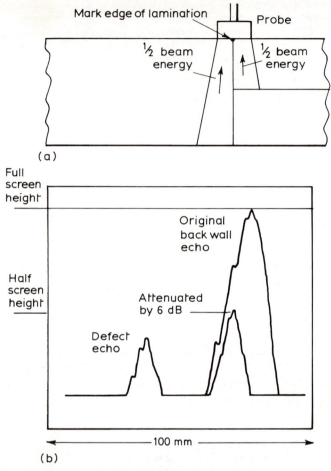

Fig. 11.28   6 dB method of plotting a lamination.

(f) Using a V1 calibration block, adjust the delay and fine range controls to achieve a minimum of four back wall echoes on the screen.

(g) Calibrate for range on the known thickness V1 calibration block.

(h) Check for linearity for time base, ensuring that the first and third back wall echoes are linear (i.e. equal distances apart).

The scanning procedure is normally specified by the client, bearing in mind the test requirements.

11.13.2   Checks for resolution

The resolution of a flaw detector is its ability to distinguish between two or more echoes at slightly differing range scales. This can be checked by placing the probe opposite the slot of the V1 block (see Fig. 11.29). In this figure, three

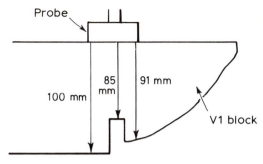

Fig. 11.29   Checking for resolution.

echoes will be received. If these are separated on the baseline, then the resolution is good.

11.13.3   Calibration of amplifier linearity

Ensure that the rejection control is set to zero. Place the probe over the 1.5 mm (0.06 in) diameter hole on the V1 calibration block. By means of the gain control, set the height of the signal to 80% of the full screen height. Then note the gain setting. Increase the gain by 2 dB (1/4 increase). The signal should now be at full screen height. Attenuate the signal to 80% of the screen height. Attenuate the signal by 6 dB. The signal should now drop to 40% of the full screen height. Attenuate the signal by a further 12 dB (12 dB = 4 : 1 ratio), and the signal should now fall to 10% of the full screen height. If the signals do not fall to the correct levels, the amplifier is non-linear.

11.13.4   Sensitivity settings

Sensitivity should be set to find the smallest specified flaw at the maximum range of the test. This normally has to be agreed by the contracting parties prior to testing.

   Generally, for compression wave scanning for laminations, the diver-inspector should adjust the gain controls until the first back wall echo (BWE) from the parent plate thickness is at full screen height. An alternative method is to adjust the first back wall echo from a horizontally 1.5 mm (0.06 in) drilled hole at the maximum range of the test, to full screen height.

11.13.5   Method of plotting lamination defect outline

When testing plate for laminations, obtain a rough assessment of what is there by a rapid scan over the whole area. Indicate any apparent flaws by marking their rough position with a chinagraph pencil. Once defects have been found, they can be sized accurately by the 6 dB method.

Using the gain control, adjust the signal to full screen height. Then move the probe towards the edge of the lamination until the signal has attenuated by 6 dB (half screen height). At this point, half the beam is being reflected by the defect and half by the back wall (see Fig. 11.28). Mark the plate at the centre of the probe to plot the lamination. This procedure is repeated all round the perimeter of the lamination until the lamination has been mapped out.

11.13.6   Sea Probe thickness meter calibration procedures

Although the ultrasonic velocity in a material is constant, the measuring system will need to be checked from time to time. This is done by comparing the reading on the instrument with the thickness of a calibration block of the material which the instrument is being set up to measure. Ideally, the instrument should be checked periodically using a step wedge so that the instrument is calibrated over the full range and not just one spot point. See Fig. 11.30.

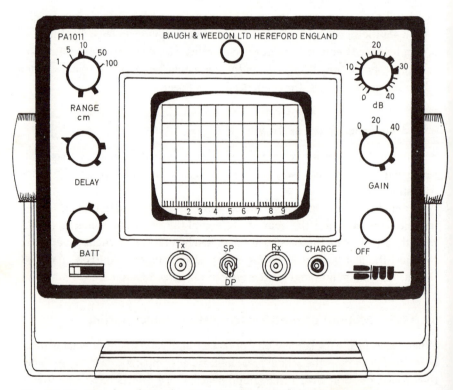

Fig. 11.30   Baugh and Weedon PA1011 ultrasonic flaw detector.

11.13.7   Sea probe thickness measurement procedures

Before taking a reading, the inspector should check the surface visually. It should be clean and free from marine fouling and rust. (There are now probes that will measure accurately through paint and protective coatings). The condition of the surface should be noted and reported and the equipment checked with a calibration block to establish confidence in the measuring system.

The first reading is taken. If it is unstable (for example, suppose that the last two figures of the digital meter keep moving), this is probably due to varying contact conditions. The instrument should therefore be removed and relocated. A little practice will soon enable the operator to obtain a consistent contact and therefore a stable reading on the meter.

Where possible, the first reading should be compared with what the thickness is likely to be. This information can be obtained from the last inspection report or from drawings of the structure. Again, this comparison will give confidence in the correct functioning of the equipment and an immediate indication as to the size of any change in thickness. Should a change in thickness be much greater than expected, it should be reported at once so that any change of programme to investigate this situation further can be carried out at once.

Should the reading indicate a change (for example, a drop to half the previous reading), the previous reading should be retaken to confirm that the instrument is still operating satisfactorily. The last reading should then be rechecked. If this lower reading is confirmed, it could be due to the material being thinner in that region, either deliberately or due to damage. Alternatively, there could be an internal interface reflecting sufficient energy to give the thickness reading only to that interface. In this case, the region should be scanned with the probe and the approximate size and shape of this region of reduced thickness should be determined.

11.13.8   Ultrasonic camera

A recent development by EMI Offshore Systems has been the underwater acoustic television system (see Fig. 11.31). This system transmits ultrasonic waves at a frequency of 2 MHz, from transponders. These waves are reflected with varying energy intensity from interfaces in their path, and are then focused by the acoustic lens system onto a pressure-sensitive plate. This plate converts the pressure of the ultrasonic signal into an electrical signal and the electrical energy of the signal corresponds to the energy of the ultrasonic signal. This is the basic information required for a picture. This picture is then fed, via an umbilical, to a television screen. As the picture is generated from ultrasonic wave information, it is independent of the light available at the inspection site. This system could be used to monitor underwater pipe laying in conditions where the stirred-up mud would preclude optical camera inspection.

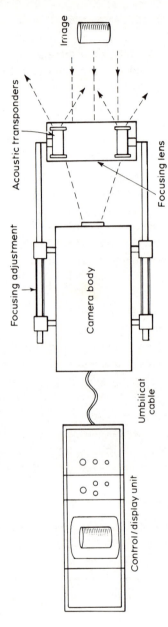

Fig. 11.31 Underwater acoustic television system, EMI Offshore Systems.

11.13.9   Gascosonic system

A more recent application of ultrasonics under water has been the introduction and use of the Gascosonic instrument. This has been designed and manufactured specifically for use under water to detect flooded members. It utilizes the stand-off method of testing where an ultrasonic signal is transmitted through the water and then into the member under test (see Fig. 11.32).

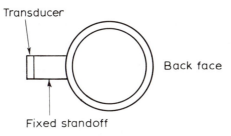

Fig. 11.32   Gascosonic system of ultrasonic testing.

If the member is intact, it will be hollow and no ultrasound can pass through to the back face. This is indicated on the CRT display by there being only one echo signal from the inside wall of the front side. This would be the normal situation. If the member has been damaged or cracked in such a way as to allow water inside, this will be indicated by the CRT display. In this case, there will be two echo signals, the first from the inside wall of the front side and the second from the inside wall of the back face. Further investigation by other test methods would then be called for in order to determine the exact cause and extent of the damage.

In use, the diver has only to place the transducer as directed from the surface. It is held in place by magnets. Once in position, there are two simple controls which are turned as directed by the surface. These two controls are used to adjust the position of the crystals in order to obtain maximum signal amplitude on the CRT, thus indicating the probe is correctly orientated and calibrated.

11.13.10   Cygnus system

Another newer instrument currently used offshore is the Cygnus handheld, single-probe, digital, ultrasonic thickness gauge. The manufacturers state that readings can be taken through paint coatings with an accuracy of plus or minus 0.1 mm (0.04 in). It works on the pulse echo method, but by incorporating a microprocessor, the manufacturers have been able to produce a meter which can indicate wall thickness in spite of paint coatings.

Finally, as with all non-destructive testing, the most efficient detection of defects depends on the skill of the operator. The instrument only assists inspectors; it never replaces them, and a background of ample experience in a technique is therefore the best investment for good inspection.

11.14   OPERATIONAL PROCEDURES

11.14.1   Typical offshore procedure when using digital thickness meters

A typical offshore procedure for this type of equipment would be as follows:

(a) The meter must be charged and maintained in accordance with the manufacturer's recommendations.
(b) Prior to each dive, the meter should be tested on a standard step which provides steps of 10, 15, 20 and 25 mm (0.4, 0.6, 0.8 and 1 in). This test should be repeated after the dive.
(c) The test site should be cleaned to a bright metal finish.
(d) Three readings should be taken on each site, ensuring that the probe is rotated between each reading.
(e) The readings should be recorded on a data sheet, together with a drawing showing the actual site of the readings.
(f) Any anomalies are to be reported immediately.

11.14.2   Method of presentation and mode of operation

When an ultrasonic flaw detector is employed offshore, the method of presentation is 'A' Scan and the mode of operation is as laid down in the current British Standards. Specific procedural requirements are as follows:

(a) The test site must be cleaned to a bright metal finish, to the inspector's satisfaction.
(b) Full details of the inspection area should be provided, for example:
  (i) Type of steel.
  (ii) Joint preparation.
  (iii) Welding process.
  (iv) Details of previous reports.
(c) Weld and joint profiles should be confirmed, where possible.
(d) Calibration prior to testing should be in accordance with current British Standards practice and should be verified on the surface.
(e) Probe characteristics must be determined and recorded prior to the inspection, and again on its completion.
(f) Any defects which are located must be sized and recorded.

(g) The results of the inspection should be reported on a data sheet, together with any drawings of location and any defects.

(h) Any discontinuities or anomalies should be reported at once.

ACKNOWLEDGEMENT

Additional material for this chapter was kindly supplied by Mr J. A. Sheppard.

12 Radiography

12.1 INTRODUCTION TO RADIOGRAPHY

Radiographic non-destructive examination of components is based on the phenomenon that materials absorb radiation energy as it passes through them. The amount of energy absorbed depends on several factors, the main ones being the density of the material and the path length in that material.

Because of their ability to penetrate solid materials, X-rays and gamma rays are used in industrial radiography in much the same way as they are used in hospitals to detect the fractures in bones.

Neutrons are also used for radiography, but at the moment they are confined to laboratory applications. The X-ray and the γ-ray are in the very high frequency, short wavelength range of the electromagnetic spectrum between ultraviolet and cosmic radiation. X-rays lie in the frequency range 10^{14} to 10^{16} kHz, that is in the wavelength band 10^{-9} to 10^{-11} m, and γ-rays lie between the frequency range 10^{16} and 1.5×10^{18} kHz in the wavelength band 10^{-11} to 10^{-13} m (see Fig. 12.1).

12.2 X-RAYS

X-rays are produced in an X-ray tube containing a tungsten filament and tungsten target (see Fig. 12.2). The tungsten filament is heated to produce a stream of electrons. These are accelerated towards the target by a high potential between the anode and cathode. This potential varies from 20 kV to 1 MV, depending on the energy requirements of the accelerated electrons.

12.3 γ-RAYS

Gamma rays are the natural radiation emitted when a substance like radium

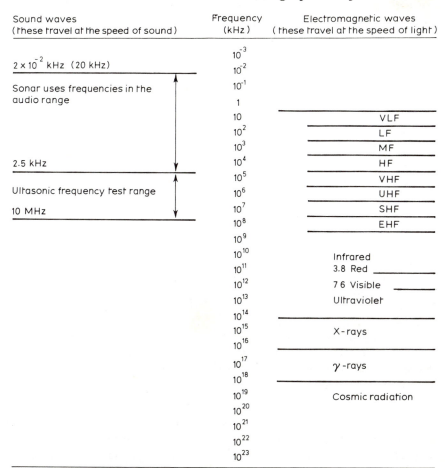

Sound waves (these travel at the speed of sound)	Frequency (kHz)	Electromagnetic waves (these travel at the speed of light)
	10^{-3}	
2×10^{-2} kHz (20 kHz)	10^{-2}	
Sonar uses frequencies in the audio range	10^{-1}	
	1	
	10	VLF
	10^2	LF
	10^3	MF
2.5 kHz	10^4	HF
	10^5	VHF
Ultasonic frequency test range	10^6	UHF
10 MHz	10^7	SHF
	10^8	EHF
	10^9	
	10^{10}	Infrared
	10^{11}	3.8 Red
	10^{12}	7 6 Visible
	10^{13}	Ultraviolet
	10^{14}	
	10^{15}	X-rays
	10^{16}	
	10^{17}	γ-rays
	10^{18}	
	10^{19}	Cosmic radiation
	10^{20}	
	10^{21}	
	10^{22}	
	10^{23}	

Velocity of a wave = frequency × wavelength
Speed of sound in air at sea level = 330 m s^{-1}
Speed of light in a vacuum = 3×10^8 m s^{-1}

Fig. 12.1   Frequencies of various types of wave.

decays, although most industrial gamma ray sources produce radiation as the isotope of a material decays (e.g. cobalt 60). The penetrating power of the radiation depends on the element from which the isotope is produced.

12.4   RADIOGRAPHIC INSPECTION

The component to be inspected is exposed to the radiation source (X-ray or γ-ray) and the radiation passes through it. The absorption of the radiation is dependent on, amongst other things, the density of the material and the radiation path length. As a rough guide, the greater the density of the material

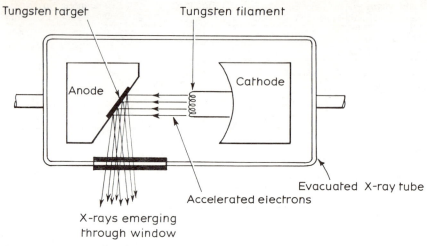

Fig. 12.2   X-ray tube.

and the longer the path length, the greater the radiation absorption (see Fig. 12.3).

In this figure, the emerging radiation I_1 has less radiation because it has travelled the greatest distance through the block that is being inspected. The emerging radiation I_2 has lost less energy than I_1 because its path length through the solid material is shorter. This is because the clear space in the centre of the block is assumed to be a less dense material, such as air.

A measure of the emerging radiation intensity is needed in order to detect any changes in it across the back surface of the item being inspected. A counter could be used, but if a permanent record is required a photographic film is more useful. It was originally found in the research into radiation that normal photographic emulsion reacted on being exposed to radiation in the same way as it did to being exposed to light. The system for radiographic inspection is therefore to place a film on the far side of the component being X-rayed (see Fig. 12.4), generally against the back face of the object being X-rayed, and then

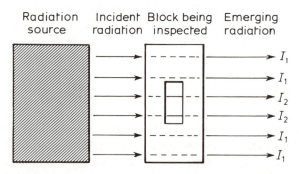

Fig. 12.3   Radiation path and absorption through a block being inspected.

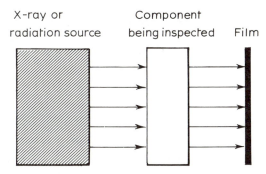

Fig. 12.4   Arrangement for obtaining a radiograph.

to expose the component to a predetermined amount of radiation. The film is then removed and developed. X-ray films usually have emulsion on both sides of the negative and there is no need to remove the negative from its lightproof packet to expose it to radiation, as the packet is practically transparent to radiation.

When developed, the negative will show varying shades of grey. The deeper the shade, the larger the amount of radiation that has fallen on the film. If we consider the block being X-rayed in Fig. 12.5, the radiograph would be darker across the centre portion where the largest amount of radiation had fallen. It must be remembered that all radiographs are negative pictures.

When depth measurements, crack sizes or maximum sensitivity of the X-ray are required, penetrameters or step wedges of the same material as that being X-rayed are included on the radiograph. Then the depth of any feature on the radiograph can be assessed either by visual comparison or by using a densitometer.

When X-rays are used for flaw detection, flaws that are parallel to the

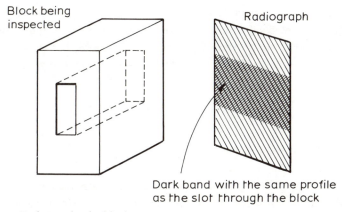

Dark band with the same profile
as the slot through the block

Fig. 12.5   Radiograph of a block.

radiation path give a much clearer definition than those that are at right angles to it because of the path length in the direction of radiation.

12.5   APPLICATION OF RADIOGRAPHY

No one type of radiation source is always used, and the choice will depend on the client's preference. In general, a radioactive isotope emitting gamma rays is convenient to handle, as it is compact and self-contained, needing no supply power cables from the surface. On the other hand, an industrial X-ray tube is often preferred both for safety reasons and for reasons of improved image.

A repair weld is invariably inspected 100% by radiography. This will be carried out in the habitat in which the welding was done, so that the environment is essentially 'dry'.

It is understood that the Royal Navy have used X-ray techniques under water without excluding the environment by the use of a habitat.

Gamma radiography is often used for the inspection of inshore docks and harbours and for concrete structures.

Pipelines are usually constructed in 12-m (39-ft) 'joints' of steel pipes. The integrity of each circumferential or butt weld must be proven by some form of non-destructive testing. Often radiography is specified as one of the tests for pipelines. This can be done in one of two ways. Either the radiation source is placed outside the pipe near the surface of the weld, or the source is sited inside the pipe. If the radiation source is outside, then the film is placed on the pipe diametrically opposite the source. Only the wall at the far side is registered, as the near wall is out of focus. There are two possible disadvantages with this method. One is that the source must be powerful enough to pass through two thicknesses of pipe, and the other is that the source must be moved around the pipe in order to cover the weld completely. Of course, this is the only possible method of inspection if there is no way of entering the pipe.

If there is access, however, an X-ray 'pig' or 'crawler' can be used. The crawler is a remotely controlled, motor driven unit carrying an X-ray tube. The motor and X-ray module are powered by integral rechargeable batteries. The film is wrapped around the weld and exposed in one shot by switching on the internal X-ray.

12.6   LAY BARGES

These are essentially seaborne pipeline construction platforms. The main method of non-destructive testing for volume type defects in the new welded pipe is X-ray. This facility is generally sited after the pretensioner and before the coat and wrap station. The technique for the radiographic examination is the same as that of a land pipe.

12.7   REEL SYSTEMS

The object of this process is to prepare the pipe on land in a spooling yard. Pipes in about 12.2-m (40-ft) lengths are welded together to form stalks of about 518.2-m (1700-ft) lengths. All the welds are X-rayed and coated. The X-ray equipment and procedure are that of the normal land-based installation.

12.8   IN-WATER RADIOGRAPHY

Currently, it is possible to take underwater radiographs using the double-wall single-image technique employing radioactive isotopes. This method of in-water radiography has been developed by C. S. Products, which is the only specialist underwater radiographic equipment supplier at the time of writing.

12.8.1   Technique when using radioactive isotopes

The isotope is secured to the worksite and the film, in a lightproof and watertight cassette, is secured on the opposite side. The source is so close to one surface that it does not register on the film, and only the far wall is recorded. Multiple exposures are required in order to achieve all-round coverage.

Divers are employed to place the source in its secured and safe condition, as well as the film cassette. When this operation is completed, the diver is withdrawn to a safe location and the source is exposed for the correct time by remote control. On completion of the exposure, the source is secured into its safe condition by remote control. Once this has been verified by remote control video camera, the diver is sent to move the source to the next exposure position. This procedure is used for each exposure until total coverage has been obtained.

13 Other methods of non-destructive testing

13.1 DYNAMIC OR VIBRATION TESTS

In its oldest form, this test involves striking the specimen and listening to the sound produced. Wheel tapping and glass ringing are examples of this in everyday use. Tests of this type involve subjecting the specimen to a single impulsive blow or a cyclic load, and measuring the response of the specimen. This response will be measured in terms of the decaying response. All other things being equal, the response of the specimen will depend on the stiffness of the specimen and the naturally inherent damping, both of which are changed by the presence of flaws. Therefore, if there were a change in response between two tests, a close investigation should be carried out to try to establish a cause. It may not, however, be due to a crack propagating; it might just be due to a change in weight.

Vibration monitoring was first installed on steel platforms in 1977 as part of an integrity monitoring programme. EMI Electronics of Woking, acting for the SEA TEK consortium, installed their system on Amoco's Montrose platform. Structural Dynamics Ltd of Southampton have used Occidental's Claymore platform and Structural Monitoring of Glasgow installed their system on BP's Forties Field.

The technique used involves analysis of the structural vibrations caused by normal wave excitation, and long-term integrity is evaluated by the changing of this pattern with time.

The system will generally employ accelerometers to detect the vibration signal. The accelerometers are mounted on the structure to a depth of 30 m (98 ft) below the surface, and the frequency analysis required to compare the vibration signature at various times will be carried out by computer.

13.2   ACOUSTIC OR STRESS WAVE EMISSION

This, in the strictest sense of the phrase, is not a non-destructive test, but it is worthy of mention as a possibility for structural integrity monitoring. It is based on the fact that an event such as the growth of a crack gives out a 'packet' of energy as it moves forward. Cracks in the early stages of growth propagate with a 'stop–go' motion. This energy is transmitted through the material of the structure as a stress wave. If these stress waves arrive at a piezoelectric crystal transducer, an electric signal will be generated. These signals can be counted electronically. From any large structure there is generally a continuous low background count, but should this count rise significantly, then further investigation of the structure should be carried out. In a test on a metal sample, this count rate starts to increase before other signs of impending failure are observed. See Fig. 13.1.

Field tests investigating the use of this technique have been carried out by the Unit Inspection Company (a part of British Steel) since September 1976, on the Dan B platform in the Danish sector of the North Sea.

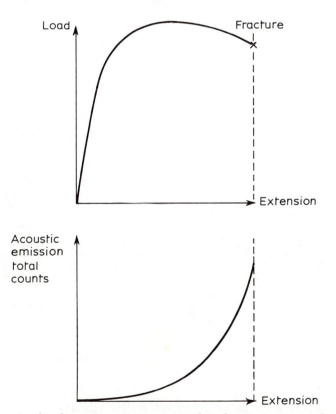

Fig. 13.1   Load and acoustic emission counts plotted against extension for a tensile test on a ductile metal.

Test	Most favourable orientation for detection	Measures thickness	Detects surface flaws	Detects sub-surface flaws	Undersea use	Material			Comments on subsea use
						Ferro-magnetic	Non-ferro-magnetic	Non-metals (concrete)	
Visual	Any	No	Yes	No	Yes	Yes	Yes	Yes	The surface must be cleaned for close inspection. Types of defect: marine fouling, scouring, missing or bent members, cracks and corrosion pitting; on concrete, spalling crumbling.
Penetrants	Any	No	Yes	No	No	Yes	Yes	Yes	
Magnetic particles	Right angles to the lines of flux	No	Yes	Only just sub-surface	Yes	Yes	No	No	Thorough cleaning required. Does not measure the depth of crack. Interpretation must be done at the time of the test and generally there is no record. (A magnetic tape method is being developed, however, and a replica of the surface can be taken.) Will detect surface cracks, seams, pits.
Radiography	Parallel to the radiation beam	Yes	Yes	Yes	Under special conditions	Yes	Yes	Yes	Thorough cleaning required. There is a size limitation on the defects detected – generally those of less than 2% of thickness are not detected. Does provide permanent record in the form of a film. Used in a habitat. Will detect internal defects, weld defects and measure thickness.
Eddy current and electromagnetic methods	Any	Yes	Yes	Yes very limited	Possible future development	Yes	Yes	No	Thorough cleaning required. Potential as a remote controlled inspection device. Defects detected: cracks, seams, pits.
Ultrasonics	Right angles to the wave direction	Yes	Yes	Yes	Yes	Yes	Yes	Yes	Thorough cleaning required. High operator skill required. Does not give a permanent record, but 'A' scan displays can be recorded. Complicated surface geometries are difficult to inspect. Will detect cracks, porosity, lamination, welding defects. The technique is sensitive and fast and can be used for sizing cracks and detecting the flooding of structural members.
Vibration or dynamic tests	—	—	Yes	Yes	Total structure in sea	Yes	Yes	Yes	
Acoustic emission	—	—	Yes	Yes	Total structure in sea	Yes	Yes	Yes	

Fig. 13.2 Non-destructive tests and comments on their subsea use.

13.3   PENETRANTS

One of the first tests used to inspect for surface flaws was the 'chalk and oil test'. The idea behind this type of test is that the presence of the crack is made more obvious by making it contrast with its surroundings.

The method used is as follows: the component to be tested is cleaned and the surface is then coated with a light-bodied paraffin. It is left for a time so that the paraffin can soak into any surface crack or flaw and the surface is then cleaned and dried. It is then coated with a chalk solution (whitewash) which is allowed to dry. The dried chalk solution, which is absorbent, leaches the paraffin out of the crack, and this produces a tell-tale stain on the white chalk many times wider than the actual line where the crack is.

The modern equivalent is based on the same principle. The paraffin is replaced by a light-bodied fluid known as the penetrant. To increase the contrast, these fluids contain a dye which is often reddish in colour, or a substance which fluoresces when exposed to ultraviolet light. The chalk solution is replaced by a fluid known as a developer, which on application dries to give an absorbent surface, which is generally white. This soaks up the penetrant that was in the crack or flaw. Both these can be applied manually with a brush or from an aerosol container.

This technique will detect only flaws that start on the surface of the material. This might at first sight appear to be a severe limitation, until one realizes that most defects (for example, cracks, corrosion pits, etc.) start from the surface of the material.

The technique cannot be used under water, but it is quite possible to use it in hyperbaric chambers and occasionally this is done. When it is used in this way, a fluorescent dye is generally used and any cracks can be viewed under ultraviolet light.

13.4   EVALUATION OF NON-DESTRUCTIVE TESTS FOR UNDERWATER USE

The table presented in Fig. 13.2 is a comment on the present state of the art, but this could well change rapidly in the next few years. The demand for more thorough and reliable non-destructive tests should see a rapid development of existing techniques as well as the growth of specialized procedures and equipment designed specifically for the offshore industry.

14 New developments in non-destructive testing

The purpose of this chapter is to outline major new developments in the methods of non-destructive testing, rather than to describe new individual items of equipment which are used in conjunction with established means of testing. These have been described elsewhere in this book.

14.1 EDDY CURRENTS AND ELECTROMAGNETIC METHODS

These are numerous, specialized tests based on these phenomena and the idea will be touched on just briefly here.

These methods are based in the main on the fact that when an alternating current flows in a test coil near or around a metal sample, it sets up eddy currents in the sample. These eddy currents produce a magnetic field which interacts with the test coil by producing a back emf (electromotive force) and changing its impedance. A flaw, change of thickness or change in the material's property will change the magnitude of the eddy current in the metal sample, and this will be shown as a change in the voltage of the test coil due to its change of impedance.

During an inspection, the test coil is passed over the surface of the component being tested. Any changes in signal from the test coil are noted and the corresponding position on the component is further inspected for a possible flaw.

Development of an unmanned submersible to inspect pipelines using this technique has been proposed and this type of inspection method might see considerable growth in the future. Thornburn Technics Inc., a British company that specializes in electronic non-destructive testing equipment, have produced an underwater version of the Halec Mark II eddy current crack detector suitable for use at depths down to 75 m (246 ft).

There is currently a diver-used item of equipment which shows promise, but this is still undergoing evaluation. The remotely operated equipment, when it is produced, will be an important addition to underwater inspection hardware.

14.2   AC IMPEDANCE

This is a relatively new addition to underwater inspection. The method is based on the voltage drop which occurs when a current is passed between two points and it relies on the 'skin effect' of alternating current.

14.2.1   Voltage drop and 'skin effect'

The voltage drop is straightforward in that the longer the current path, the greater the voltage drop. The actual voltage drop is dependent on the resistance of the material. For example, for copper it is relatively small, but for constantan it is relatively high. The 'skin effect' describes the way in which an alternating current travels along a conductor or current path. The current travels along the outer skin rather than through the centre of the conductor.

These two effects are utilized in the following manner. A current is passed at right angles to the line of the flaw so as to bridge the flaw. The voltage is then monitored on sound material and then across the flaw. The two readings are then compared and the difference between the two is an indication of crack depth. See Fig. 14.1.

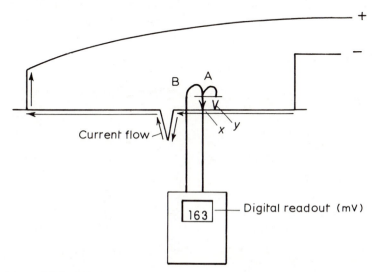

Fig. 14.1   AC impedance.

A reading is taken at A and again across the crack at B. The voltage path between the prod tips (x and y) is greater at B, and therefore the voltage drop is greater.

14.3   FIBRE OPTICS, OR ENDOPROBES

Visual observations can be improved and visual inspection of inaccessible sites, such as the inside of gas turbines, can be made possible by the use of endoprobes. The flexible version can see around corners.

The system consists basically of fibre light guide cables, a light source and an optical system. The field of view can be varied from forward through lateral to retroview, of the order of $135°$. The lenses and optics are usually colour corrected to give accurate reproduction. Reflex cameras, Polaroid cameras and television monitors can be used for recording purposes. Gratticules can be fitted to make possible visual assessment of size, and are especially useful when investigating pits, cracks and areas of corrosion. A pit depth gauge can also be fitted, measuring pit depth by the use of a micrometer mechanism.

14.4   INFRARED OR THERMAL IMAGING

This technique has been developed for land-based inspection and works on the principle that the presence of defects in a material will change very locally the thermal conductivity.

The component to be inspected is exposed to infrared radiation (or a laser) and the initial rise in temperature of the surface is viewed by the thermal imaging camera. If there is a defect present, then the rate of heat transfer in that area changes and the infrared camera will detect changes of temperature, thus indicating the site of the defect.

15 Reporting and documentation

15.1 THE NEED FOR REPORTS

All the structures standing in the North Sea are subjected to considerable deteriorating forces, as outlined in Chapter 6. Of course, the designers take this into account when designing the structure, but local anomalies do occur and some detailed aspects of the problem are imperfectly understood. Another point to note is that any large structure of the type seen offshore warrants careful monitoring on engineering grounds.

This indicates a need for documentation for offshore structures and the importance of these records should not be underestimated. The average working life of an offshore structure was initially predicted to be some 25 years. During that life cycle, it would be reasonable to assume that for reasons we have outlined earlier defects of one type or another will occur. It makes good sense for both engineering and economic reasons for any such damage to be dealt with on a planned basis.

The importance of documentation and records can be looked at under several headings as follows.

15.1.1 At the fabrication stage

At this stage, any allowable defects must be recorded. Then, if any problems occur during the working life of the structure, those fabrication defects can be considered within the overall picture. The effects of any local increase in stress cannot be properly calculated unless these defects are documented.

15.1.2 During the installation of the platform

During this stage, extra stresses are imposed on the structure. It is probable that the level of these stresses will not be reached again during the remaining life of that structure. If any failures or defects occur at this point, it is vital that they are recorded so that engineers can remedy or monitor them. Remedial action or monitoring will be dictated by the site, magnitude and overall stress path for the defect under scrutiny.

15.1.3   In-service stresses

These are caused by:

(a) Environmental forces.
(b) Possibly by overloading.
(c) Accidental damage.

Any damage caused by these factors must be recorded with the appropriate engineering action also being logged. This builds up a total picture of the structure. Maintaining these records makes it possible to ensure a fail-safe working life for the structure while maintaining a highly cost-effective repair and maintenance programme.

On any offshore platform, there are men working all day every day. Should the platform fail, then there exists the possibility of many lives being lost. There is a high risk of contamination of the sea with crude oil and the possible environmental repercussions of this to consider also. Finally, any failure of an offshore structure would be extremely expensive because of the loss of income and because of the cost of rectification or repair. This underlines again the importance of documentation to ensure effective maintenance of all offshore structures.

15.2   THE INFORMATION REQUIRED

When a report is collated from information gathered from visual inspection, the information must include the following.

15.2.1   Type of damage

This will be detailed as accurately as possible with a full description, if required. At present, there are no standard terms for damage other than BS 499, which deals with welding faults. This can make the task of describing the damage that much more difficult in some cases. Generally, however, terms such as 'dents', 'scrapes', 'scratches' and 'bows' are precise enough and are widely used.

15.2.2   Position of damage

The position of the damage will have to be fixed and this will usually be in accordance with the client's instructions. On circular members, the identification letters are used to identify the component and the clock method is used to indicate position.

15.2.3   Reporting cracks

When cracks are reported, as well as the length and width, the orientation of the crack is also usually required. When it is possible to determine the rate of increase or decrease in length and width, this is also recorded. The direction of crack propagation and whether the crack is on the weld, in the HAZ (heat affected zone) or on the parent metal, is also required.

15.2.4   Magnetic particle inspection

When a magnetic particle inspection is undertaken, the following information will be required in the report, where a standard procedure and prepared technique sheets are employed:

(a) Work location.
(b) Description and identity of the components tested.
(c) Date of test.
(d) Stage of test (e.g. after visual inspection).
(e) Reference to the written test procedure and the technique sheet used.
(f) Name of the company.
(g) Name and signature of the person conducting the tests.

Where no procedure sheet is used, the following information will be required in an MPI report:

(a) Description of equipment used.
(b) Technique of flux generation.
(c) Indicated current values and wave form for each technique.
(d) Spacing on prods or details of coils.
(e) Details of inks.
(f) Surface preparation.
(g) Viewing conditions.
(h) Method of recording.

15.3   METHODS OF REPORTING

These fall into two major categories at present: written reports and computer recordings. In both cases, there are reporting procedures to adhere to and the following paragraphs outline the requirements for each.

15.3.1   Reporting procedures

In some cases, the reporting procedure is laid down in detail by the client. In these cases, the size of paper, the format, the paragraph numbering, the sequence and even the information required are all detailed in instructions issued by the client.

In other instances, the procedure for the report is not so carefully detailed and it then becomes the responsibility of the contractor. Individual inspectors are not required to evolve their own reporting procedures. They must ensure that they use the procedures on a particular contract correctly and that important inspection information is presented for analysis.

15.3.2   Written reports

It is not possible to lay down hard and fast rules for written reports, as these vary from one company to another. The following guidelines, however, are generally correct for most occasions.

(a)   Format of the report

The report should be typed on company headed paper, and bound. It should have a unique report code reference and latest issue number on each page, as well as the date of the report. All pages should be numbered, and the first page should be an amendment sheet, if necessary, showing the latest amendments since the original report was formulated. The report should have a specific and relevant title, as well as a table of contents, and the sections and paragraphs within the report should be accorded numbers and sub-section numbers. Drawings may be required to show the site, the actual component and any defects. These must also be numbered. Also, any relevant photographs must be included in the report and be cross referenced, as should any relevant video tapes. Lastly, the report must be approved and signed by a competent person within the company.

(b)   Sections within the report

(i) *Scope*   This should outline what the report is for, e.g. 'This report describes the technique for the Magnetic Particle Inspection of node joints (here the part numbers and other means of identification of the components under test, including their location, should be given) using flexible coil'.

(ii) Relevant documents   All documents which are referred to in the report should be listed in a standard fashion, with their title, date and origin. For example, 'BS 6072: 1981 – Methods for Magnetic Particle Flaw Detection. British Standards Institution'.

(iii) Personnel   Details of the personnel who carried out the test must be carefully stated. Either the company will have a written practice for training and certification of personnel, or else other qualified operators should be employed. It is not sufficient to write, for example, 'CEGB'. It should be written as '3.1 CEGB MT Level II, Category 2 (in accordance with document 989904)'. Do not include qualifications which are irrelevant.

(iv) Equipment used   All equipment must be listed. Control checks and equipment calibration should be given in a separate section on Calibration, and the equipment list should include some form of reference check, e.g. MT – Penetrameter. Where consumables have been used, trade names should be used.

(v) Surface condition   The surface condition of the component under test and its preparation prior to test must be stated, e.g. machined to 125 CLA maximum.

(vi) Method of testing   This will be one of the most detailed sections of the report, as it will describe precisely how the test has been carried out.

(vii) Results of testing   Here again, a detailed description of the results of the test is essential, giving the size and nature of indications of the test and the date of the test. All details should be written in accordance with the customer's requirements and all relevant indications must be considered with regard to the acceptance criteria of the procedure document. Where the report is used to give a verdict on the acceptability of the component, all relevant criteria must be included in the report. At the very least, reference should be made to a formal document containing acceptance limits, e.g. the customer's order or standards. Where no such acceptance levels are specified, which is usual in some underwater inspections, and the diver-inspector is simply reporting his/her findings, this fact should be recorded in the report. Any drawings, photographs, etc. which are relevant should also be included in this section, together with references to those items and/or any videos used in the test.

(viii) Status marking   This section should provide details of the way in which parts of the component have been marked. For example, red for unacceptable, white for acceptable, and blue for those parts awaiting further examination.

Thus, the report should include as much relevant information as possible in a clear and precise form.

15.3.3   Video recordings and their use in reporting

Video recordings can be made on different kinds of equipment using different formats. The most generally used is the 19 mm (3/4 in) U-matic format. Frequently, video typewriters are employed and usually a timer is used to show time and date. The diver's depth and even CP readings can also be displayed on the screen. Colour is used more often now and monochrome cameras are on the decline.

A voice commentary is often required, and it must cover the following:

(a) Details of the actual area being viewed.
(b) The reasons for the survey.
(c) Where relevant, the visibility, current and seabed conditions.

The commentary should be given in a clear conversational voice. It must be comprehensive, give regular reminders of location, and give reasons for any pauses, relocations or interruptions.

The video recording normally should be identified by the job number, structure, component number and dive number, as appropriate.

The video recording is primarily used to undertake structural surveys, pipeline surveys (by submersible or ROV), and component inspections.

15.3.4   Computer recordings

These are quite sophisticated and consist of written reports and drawings. The raw data are fed into the computer by using standard forms which follow the programme layout. Drawings are prepared in draft form and these too are then fed into the computer for storage. Currently, all the information is duplicated and is stored as 'hard copy' in case the computer 'dumps' any part of the report.

The actual information in the computer is stored on discs which can be despatched to any location as required, and they can also be copied if necessary. The great advantage of computer recording is the ease with which information can be accessed. This means that interpretation of the inspection can be both quicker and more detailed.

15.4   SPECIMEN REPORTS

The following figures show brief examples of specimen reports (Figs 15.1–15.3).

WELD No:		BP WEST SOLE FIELD UNDERWATER INSPECTION	REPORT REF:
WELD REF:		GENERAL WELDMENT INSPECTION DATA SHEET	DATE:
PLATFORM:	DATA RECORDER:	DIVER:	DIVE No:

MARINE GROWTH

	HARD	SOFT
TYPE:		
THICKNESS:		
% COVER		

% WELD ACCESSIBLE:

PHOTO REF:

VIDEO REF:

WELD LOCATION SKETCH:

INSPECTION CO-ORDINATOR:

VISUAL INSPECTION

TOPCOAT:

PRIMER:

BSM:

BLISTERS:

DEPOSITS:

CORROSION:

DEFECTS:

DAMAGE:

CV & MPI REPORT REF:

BP REP:

SPECIMEN

Fig. 15.1 Inspection report sheet.

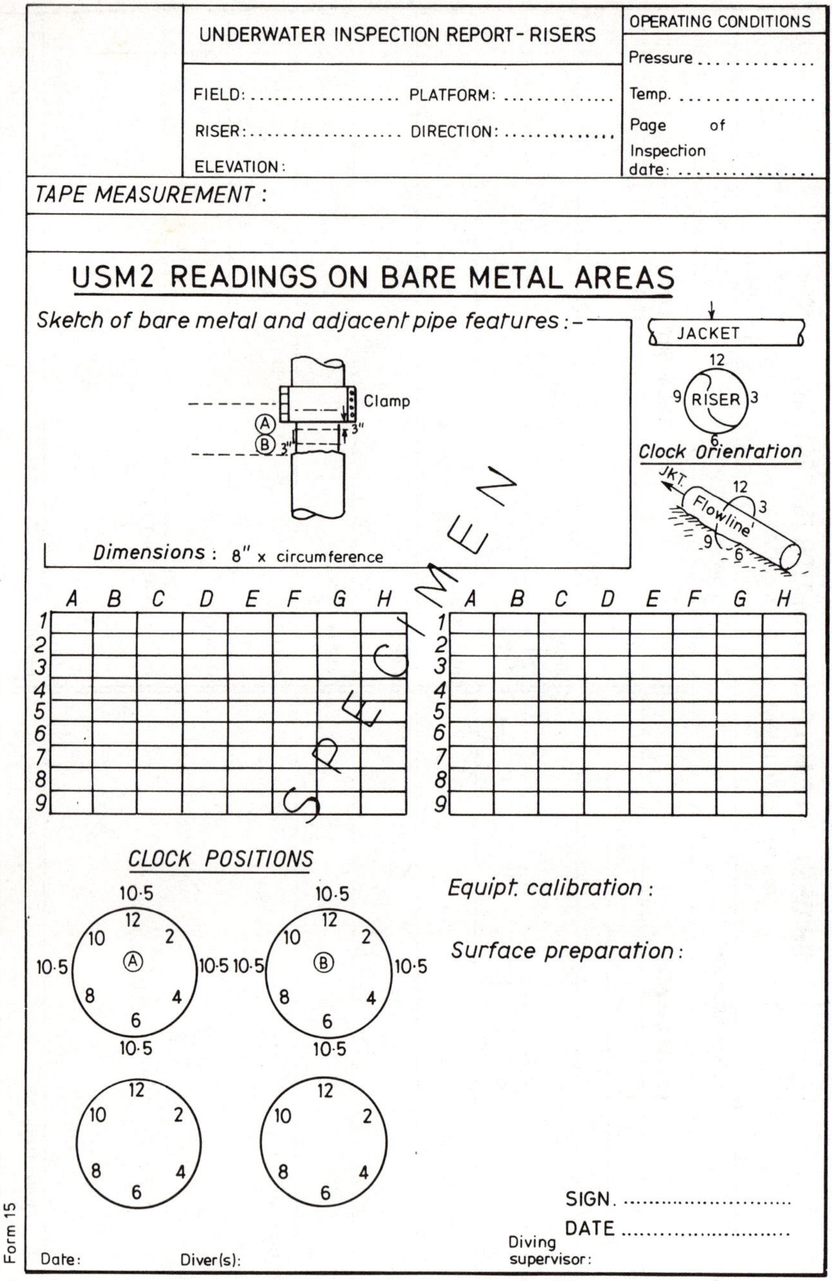

UNDERWATER INSPECTION REPORT - RISERS

FIELD: .... PLATFORM: ....

RISER: .... DIRECTION: ....

ELEVATION:

OPERATING CONDITIONS

Pressure ....

Temp. ....

Page of

Inspection date: ....

TAPE MEASUREMENT :

USM2 READINGS ON BARE METAL AREAS

Sketch of bare metal and adjacent pipe features :–

Dimensions : 8" x circumference

SPECIMEN

	A	B	C	D	E	F	G	H
1								
2								
3								
4								
5								
6								
7								
8								
9								

	A	B	C	D	E	F	G	H
1								
2								
3								
4								
5								
6								
7								
8								
9								

CLOCK POSITIONS

Equipt. calibration :

Surface preparation :

SIGN. ....

DATE ....

Form 15

Date: Diver(s): Diving supervisor:

Fig. 15.2 Inspection report sheet.

VIDEO TAPE LOG

Diver ------ Client ------

Date ------ Contract ------

Location ------ Underwater visibility ------

Component ------

Type of inspection ------

Reel cassette No.	Time counter No.	Job description	Location	Remarks

SPECIMEN

Fig. 15.3 Inspection report sheet.

16 Repair of offshore structures

16.1 REPAIR

The major repairs to offshore structures seem to fall into two main groups: the repair of steel structures, which are mainly tubular in section, and the repair of concrete structures.

The incidence of major repairs undertaken has steadily increased from just two major repairs out of a total of 52 platforms in 1973, to 21 major repairs out of a total 120 platforms in 1981, in the North Sea alone. A majority of these were carried out above a depth of 20 m (66 ft).[1]

The principal causes of the damage that necessitates repairs are given below. Surprisingly, although seawater is highly corrosive, corrosion has not been a major problem in itself. However, it must have accelerated corrosion fatigue damage.

The three main causes of repair are:

(a) Design or construction faults.
(b) Fatigue.
(c) Impact damage.

16.1.1 Design or construction faults

In the case of design or construction faults, these have been due mainly to ignorance of the environmental and loading conditions required to be withstood by the structure. As expected, there is a higher incidence of damage requiring repair on the earlier platforms.

16.1.2 Fatigue

Fatigue cracking needing repair has been observed on structures. The main areas at risk are the nodes, because the understanding of the design and fabrication of these complex load-transference structures is still imperfect, and conductor bracings in comparatively shallow water have experienced fatigue problems caused mainly by the wave loadings.

16.1.3   Impact damage

Impact damage is probably the most common cause of structural damage, and it mainly occurs in the first 30 m (98 ft) below sea level.

The main methods of repair to steel structures are welding, friction clamps and grouted clamps. Welding is most often used, as it is fast and cost-effective and requires much less precise information than that required for clamps. Careful design and execution of hyperbaric welding can return the structure to a better than original condition.

For concrete structures, the nature of the repair will depend on the type of concrete structure and the extent of the damage. The repair task may therefore involve the repair of the reinforcement or prestressing tendons, prior to the repair of the concrete itself.

16.2   DIVING INVOLVEMENT

The diving system that is needed to repair damage below the waterline will be determined by a complex interaction of many factors such as the accessibility of the damaged area, the nature of the repair that is required to restore the integrity of the structure, as well as the cost and availability of the diving facilities. The type of diving facility used might be platform-based or a large, small or semi-submersible diving support vessel. At the work site, the diver will be required to work in the water in a free-swimming mode, in an atmospheric caisson, in a wet habitat, or in a dry habitat.

Before the repair can be carried out, an accurate assessment of the extent of the damage must be carried out. This will certainly involve the cleaning of the marine growth from the steel node or surrounding concrete structure by water jet or some other mechanical tool. A water jet will also be used to remove loose concrete and to clean the concrete away from the ends of the reinforcing rods. A thermic lance can then be used to cut away the reinforcing rods or damaged steel member. At the end of this phase of preparation, the diving operation should be capable of supplying pictures of the site to be repaired with an accurate profile of the dimensions required in order to design the repair structure and the procedure for its completion.

16.3   WET WELDING

This technique has been in use for many years and has been demonstrated as being reasonably effective during that time.

The basic technique is arc welding, where a DC generator provides power to an electrode and an earth clamp. The clamp is secured to the workpiece and the electrode consists of a mild steel rod coated with a flux covering which has a waterproof covering over the top. As soon as the electrode comes in contact with the workpiece, the circuit is complete and the welding continues as normal. There are numerous problems associated with wet welding, including the following.

16.3.1   Quenching

In this case, the water causes the weld to cool too quickly and also prevents preheat treatment. During welding, higher current has to be used to compensate for the quenching effect. This can cause undercutting.

16.3.2   Hydrogen embrittlement

Here, the high energy arc can break down the water molecules and allow hydrogen to percolate into the weld pool.

16.3.3   Lack of fusion

In multipass welds, there may be lack of fusion between passes because of the problems of trying to maintain interpass temperatures.

Because of these types of problem, wet welding was frowned upon in the North Sea until 1985. Its only use was in temporary repairs or in applications such as welding on ancillary anodes. Where structural repairs were required, hyperbaric welding was used. During 1985, however, several wet weld repairs were undertaken in the North Sea using ASME (American Society of Mechanical Engineers) specifications. These welds were not in highly stressed areas, but none the less, they were structural and they were undertaken actually in the water and not in a hyperbaric chamber. They were acceptable to the certifying authorities and it can be expected that this type of welding will become acceptable throughout the North Sea in the near future.

16.4   HYPERBARIC WELDING IN DEEP AND SHALLOW WATER

This form of welding is currently the main type of welding employed in the North Sea that is certifiable by the authorities.

The method entails taking a chamber to the site for the repair, and installing it in position with watertight seals. Once the seals are in place, any water inside the chamber is evacuated by increasing the internal gas pressure within the chamber. The gas mixture in use is oxy-helium, so that saturation techniques can be employed for the welder divers.

The weld preparation, technique, consumables, pre-heat, post-heat and inspection requirements are all laid out in the welding procedure. The hyperbaric chamber is kept on site until such time as the final inspection requirements have been met. The seals are then broken and the chamber is removed to the surface.

One of the preferred welding techniques is to use tungsten inert gas (TIG) welding. This method gives good penetration into the root and generally does not suffer from problems of porosity or lack of fusion.

The welders are usually specially selected for this type of work and the normal method of working is to have divers do all the preparation such as cutting away any unwanted material, placing the seals, placing the chamber and any associated tasks. The welders then enter the hyperbaric chamber for the final weld preparations and the welding itself.

During the welding operations, the supervisors on the surface have a very demanding job. The atmosphere in the chamber must be monitored very carefully in order to ensure that no contamination builds up during the welding process. The wellbeing of the welders, as well as that of the rest of the diving team, depends entirely on the supervisors.

Once the hyperbaric chamber is in place and the welders are working, it is no simple task to recover either the welders or the chamber, and this requires close supervision in order to avoid any problems.

The type of hyperbaric welding described so far applies to that where oxy-helium saturation techniques are employed. This method is usually used below depths of 30 m (98 ft). At depths of less than 30 m (98 ft), air can be used, and the welding can be undertaken using air saturation techniques. Thus, the welding techniques are the same whether the welding process takes place in shallow or deep depths; the main difference is simply the breathing mixture used. The main difficulty with shallow water welding is that it can be much more troublesome to install the chamber itself because of tidal and wind effects.

16.5   CEMENT GROUT INJECTION OF THE CHORD

Grouting the chord of a tubular connection can improve its fatigue life, as the grout will stiffen the chord wall and therefore reduce the stress concentration at the joint. However, if the joint has already cracked, the effectiveness of this repair is in question. In some instances, the grouting may have an adverse effect, since the joint bending stiffness will increase, which may attract an increase in the bending moment and therefore the stress at the joint. Careful analysis must be undertaken before this method is used. This method is used to strengthen a structure rather than repair damage to the structure.

16.6 BOLTED FRICTION CLAMPS

These can be used to fix risers, casings and other utilities to steel structures, but they are generally avoided for main structural connections. The clamps must form a good fit with the tubular member, and hence the section on which the clamp is to be fixed must be cleaned to obtain a good surface finish. The section must then be accurately measured. See Fig. 16.1.

The tightening stress of the bolts has to be carefully calculated and applied in order to avoid crushing the member. The bolts on friction clamps have to be checked and replaced at intervals during the life of the clamp, as the fluctuation of loads in the inner member leads to fluctuating bolt loads superimposed on its initial pretension, causing the bolts to stretch or work loose.

This type of technique has been adopted for the repair of fatigue cracking on the nodes in the conductor guide frame. Strengthening is achieved by wrapping formed plates around the main node member. The two-piece chord wrap member is then welded in place.

16.7 GROUTED CLAMPS

Repair of nodes is probably the most critical use of grouted clamps, but they are also used for strengthening or replacement of individual members or the installation of additional braces.

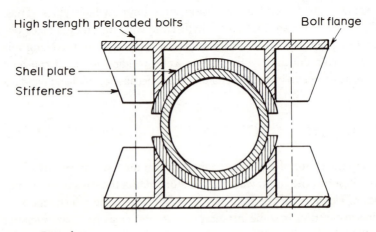

Fig. 16.1   Pipe clamp.

Simply, the clamp is made oversize and is then fitted over the joint. Rubber or inflatable grout bag seals are installed and the annulus grouted. After curing, the clamp bolts are preloaded. The use of liquid grout removes the necessity for really precise tolerances between clamp and workpiece. Stressed grouted clamps have significantly greater load capacity than ordinary friction clamps.

16.8   REPAIRS TO REINFORCED CONCRETE STRUCTURES

The repair to reinforced concrete structures falls into two parts. First of all, the damage to the steel reinforcement or prestressing tendons must be made good, and then the concrete cover must be reinstated.

Repair to the steel reinforcement can be carried out by a variety of methods, namely:

(a) Lap joints
(b) Welding
(c) Mechanical couplers
(d) Grouted connections
(e) Steel plates.

These all provide continuity of reinforcement from the damaged ends of the steel reinforcement.

Repair to the prestressing tendons has to be carried out in such a way that not only is the tendon prestressed, but also the new concrete used in the repair is stressed to prevent it from cracking.

The concrete cover can be repaired with either concrete or resin. Generally, resins are used to seal cracks and fill small holes in order to stop the ingress of water.

The differences in behaviour that make resins suitable for the repair of concrete structures are their high compressive strength (higher than concrete), their low water absorption (lower than concrete), the fact that curing can be very quick so that the strength is developed very rapidly, and the fact that, as with most polymers, they have good chemical resistance. However, on curing, the exotherm can give quite high temperature rises and so the formulation of the polymer needs to be accurately monitored. If unfilled, they are prone to creep.

Concrete is used for repair to large areas of concrete damage, and has the advantage of being cheaper than resin repairs.

Whatever the type of replacement material used, the properties achieved will depend on its exact application. It is essential that whatever material is used, it will be suitable to be placed under water and form a good bond to existing concrete and reinforcement.

16.9   INSPECTION REQUIREMENTS FOR REPAIRS

The first point to be made here is that the inspection procedure itself is inspected. This inspection takes the form of a review in order to ensure that the procedure is correct and that the proposed repair will be fit for its task. This then validates the repair procedure and the next requirement is close supervision in order to ensure that the procedure is followed carefully.

Finally, the actual repair itself is inspected using the same techniques and criteria as employed for fabrication and in-service inspections. This part of the inspection may be undertaken by different personnel from those who actually undertook the repair, in order to maintain impartiality. The same standards apply to post-repair inspections and reports as apply to the original fabrication inspection reports.

16.10   CASE HISTORIES

16.10.1   Cases of damage

The following is a list of some of the accidents that have occurred in the offshore industry:

Year	Accident
1965	Sea Gem collapsed into the North Sea while drilling proceeded in hostile conditions.
1975	Howards jack-up drilling barge lost one of its 36-m (118-ft) legs in a storm. The structure had suffered a history of fatigue cracking of welds.
1975	Ekofisk. Corrosion damage to a pipeline caused a pipe to rupture and spray hydrocarbons over the living quarters.
1975	Frigg's DP1 failed before installation. Incorrect design of the flotation tanks caused it to sink in 107 m (351 ft) of water. It was later salvaged.
1975	Brent 'A'. Jacket completion was delayed when a brace failed.
1975	In the Shell–Esso Auk field, a supply boat rammed a steel production platform.
1976	The British tanker *Globic Sun* collided with an unmanned oil platform in the Gulf of Mexico.
1977	There was a fire on Ekofisk Bravo.
1978	Statfjord 'A'. Five men working inside a concrete leg were killed by fire and smoke.

1979   Ranger One. This three-leg jack-up collapsed in the Gulf of Mexico. The cause was long-term fatigue. Two fatigue cracks, 320 mm (12.6 in) and 450 mm (17.7 in) long were found at the junction of the aft leg and steel seabed mat.

1980   The Alexander Kielland, a five-leg semi-submersible suffered a leg severance. The cause was a materials failure by fatigue of major proportions, which had gone unnoticed before fracture.

1981   The Cabinda Gulf rig, located off Angola, collapsed into the sea. The cause was thought to be gas leaks weakening the seabed.

1983   Key Biscayne. A jack-up platform was lost off Australia, 120 miles (193 km) north of Perth. As there is no plan to recover the platform, the cause of failure is still a matter of speculation.

1985   The North Star jack-up rig was damaged in heavy seas off South Africa, causing partial collapse.

16.10.2   Principal causes of accidents

The principal causes of accidents could be classified in the following way:

(a) Collision. Planes, ships, oil rigs.
(b) Blow-outs. Environmental disasters.
(c) Capsizing, due to instability, especially in rough seas or due to the development of seabed instabilities.
(d) Major mechanical failures, where a major part of the structure becomes inoperative. Fatigue cracking has been the major cause in most of these events.
(e) Minor mechanical failures. The major cause of minor damage is impact by vessels colliding with the structure or by objects being dropped during installation or unloading supplies. This damage is predominantly in the first 30 m (98 ft) below sea level.
(f) Grounding of towed or runaway rigs and barges.

Minor mechanical failures are the most usual to involve repair by divers on site. The following are brief descriptions of repairs carried out by Oceaneering.

Fatigue cracking problems were encountered on the K nodes in the conductor guide frame of a structure at 5 m (16 ft) depth. The remedial action was to strengthen the nodes by wrapping formed plates around the main member and welding them. The regions most at risk from fatigue were dressed and examined by ultrasonics, and all welds were subjected to toe grinding and inspected using MPI.

The repair of boat impact damage on a North Sea gas platform was effected by replacing a diagonal brace between +5 m (16 ft) and −6 m (−20 ft) elevation, and the horizontal brace at −6 m (−20 ft). Telescopic replacement braces were used and cracks in associated nodes were repaired by welding.

REFERENCE

1. UEG (1983) *Repairs to North Sea Offshore Structures – A Review*, Report UR 21, Underwater Engineering Group, London.

Index

Entries in italics refer to figures